SpringerBriefs in Electrical
and Computer Engineering

For further volumes:
http://www.springer.com/series/10059

Costas Laoudias · Costas Psychalinos

Integrated Filters for Short Range Wireless and Biomedical Applications

 Springer

Costas Laoudias
University of Patras
Rio Patras, Greece
laoudiask@upatras.gr

Costas Psychalinos
University of Patras
Rio Patras, Greece
cpsychal@physics.upatras.gr

ISSN 2191-8112   e-ISSN 2191-8120
ISBN 978-1-4614-0259-6   e-ISBN 978-1-4614-0260-2
DOI 10.1007/978-1-4614-0260-2
Springer New York Dordrecht Heidelberg London

Library of Congress Control Number: 2011937579

Printed on acid-free paper

Springer is part of Springer Science+Business Media (www.springer.com)

Preface

The technological evolution and market requirements have led to an increasing demand of low-power portable devices, featuring the reduced size and high efficiency. The supply voltage scaling sets new challenges in analog circuit design as this leads to a degraded circuit performance in terms of available bandwidth and smaller available signal swings. So, the design of high performance circuits using current mirrors is highly affected from these parameters. In this direction, novel analog integrated filters using low-voltage high swing current mirrors are introduced. The resonant frequency of the proposed filters can be electronically controlled by modifying appropriate dc currents. Thus, any effect from process, voltage and temperature variations can be eliminated. The derived topologies can be applied in low-power implantable devices for biomedical applications and electronic devices for short-range wireless communication systems.

This book covers theoretical aspects of low-voltage current mirrors and the realization of integrators' topologies using this active cell. A systematic study in the design of high-order filters using several methods is also presented, thus providing an overview in the design of analog filters using current mirrors. Several design examples with the test setup and extended experimental measurements are included in order to demonstrate the performance of the presented topologies. Much of the material presented in this manuscript, originates in work done by Costas Laoudias through his Ph.D. research at University of Patras, Greece, supported by the Research Committee of University of Patras under the Karatheodori Project (C.158).

The authors would like to thank Prof. Mohammed Ismail for his valuable comments and his encourage writing this book. Finally, many thanks to all the members of the Electronics Laboratory of Physics Department, University of Patras for their helpful efforts and stimulating technical discussions over these years.

Rio Patras, Greece

Costas Laoudias
Costas Psychalinos

Contents

Chapter 1
Introduction

1.1 Low-Voltage Analog Circuits

Advances in VLSI technology combined with increase market demands to develop efficient portable devices have increased the interest in designing circuits that are capable to operate at low-voltage and/or low-power consumption. The design of analog and mixed-signal integrated circuits that can operate in a low-voltage environment with high performance is now imperative. This mainly stems from the continuous shrinking of device sizes, which leads to lower breakdown voltages and thus, in fact, circuits cannot operate with high voltages. Another reason is that modern applications require handheld devices with as small as possible dimensions and increased longevity, like the implantable pacemakers for the detection of cardiac signals and radio devices for short-range wireless communications. Also, co-integration of analog and digital circuits on the same chip, as required in modern mixed-signal system-on-chips (SoCs), implies that analog circuits must operate from supply voltages as low as digital ones. The International Technology Roadmap for Semiconductors (ITRS) [1] gives us a unique opportunity to look into the projected future of semiconductor technology and identify the design challenges. According to ITRS, by about the year 2013 and at the 32 nm technology node, the power supply for digital circuits will be at 0.5 V. Since the power consumption in digital circuits is proportional to the square of supply voltage, the continuous voltage scaling results in the reduction of overall power consumption. Of course, it should be noted that the reduction of supply voltage is mainly determined by the reliability of circuits, breakdown voltages and thermal problems; thus, they have to be appropriately scaled down.

On the other hand, in analog circuit design, the supply voltage scaling results in the reduction of the available signal swing, and therefore in less dynamic range for a given power consumption. As a consequence, the elimination of circuit errors due to thermal noise or offset voltages (currents) eventually leads to higher power consumption [2–4]. Thus, the existing analog circuits should either be modified or redesigned using new circuit design techniques in order to have the same or even better performance under low-voltage conditions. These low power supply voltages

C. Laoudias and C. Psychalinos, *Integrated Filters for Short Range Wireless and Biomedical Applications*, SpringerBriefs in Electrical and Computer Engineering, DOI 10.1007/978-1-4614-0260-2_1, © Springer Science+Business Media, LLC 2012

(1.5 V and lower) and the relatively high device threshold voltages (about 0.5 V at 0.35 μm process) are the major obstacle in the design of high performance analog circuits. Moreover, the supply voltage constraint sets a limit on the number of transistors that can be stacked, since one part is used for the bias circuit and the remainder must be available for the signal swings.

A term to define an analog circuit as low-voltage is the sum of gate-source (V_{GS}) and drain-source (V_{DS}) voltages of the MOS transistors that are connected between the supply voltages. It is well known that, the minimum supply voltage required to operate an MOS transistor in saturation region is typically determined by two parameters, namely the threshold voltage V_{th} and the saturation voltage $V_{DS,sat} = V_{GS} - V_{th}$. Therefore, low-voltage operation is achieved by utilizing several circuit design techniques that are minimizing these two parameters. The threshold voltage can be lowered by modifying the process and using low threshold voltage or native zero threshold voltage transistors. However, technologies with low-threshold transistors are often available at the expense of higher production cost. Since analog circuits occupy only 5–30% of the total die area on large SoC circuits, then this solution is inappropriate. Another way to reduce the threshold voltage is by using design techniques that have been proposed in the literature, such as floating and quasi-floating gate transistors, self-cascode transistors, bulk-driven transistors etc. Nevertheless, in most cases, V_{th} has a fixed value and is not a design parameter that can be adjusted in some way. On the other hand, the saturation voltage is under the designer control. For instance, by minimizing the $V_{DS,sat}$ while using very small bias currents, the MOS transistors end up operating either in triode or subthreshold region. Thus, the usage of transistors in these operation conditions is the most common technique in low-voltage/low-power application.

Recently, another approach that is used for the design of low-voltage circuits is the current-mode technique [5]. This subclass of circuits is the adjoint of voltage-mode circuits, where input–output signals as well as intermediate output signals are currents. An attractive characteristic of current-mode circuits is that they present low-impedance internal nodes. The key performance feature of current-mode circuits is their inherent potential to operate at higher signal bandwidths than those achieved by conventional voltage amplifiers. This stems from the fact that their active building blocks do not present high gain while the time constants formed by the parasitic capacitances and low-impedance of internal nodes are very small. Finally, current-mode signal processing is a very attractive approach for many applications due to the simplicity in implementing operations such as addition/subtraction, multiplication by a constant, and inversion.

1.2   Low-Voltage Current Mirrors

One of the most simple and reliable building block for performing current-mode signal processing are the current mirrors. They are a very common active element both in analog and mixed mode VLSI circuits and are used either for performing

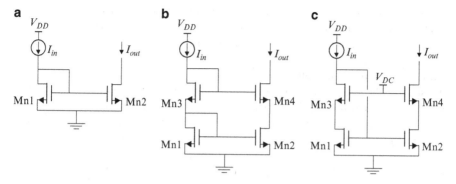

Fig. 1.1 (a) Simple current mirror (b) conventional cascode current mirror (c) low-voltage high-swing cascode current mirror

signal processing (i.e. continuous-time filters) or biasing. As mentioned in the previous section, the reduction of supply voltage and transistor dimensions as well as the effect of channel length modulation set new challenges in the design of analog circuits. As a result, the design of high efficiency low-voltage low-power current mirrors is also affected. Several topologies of current mirrors have been proposed in the literature, which are aimed at improving characteristics such as the input/output impedance, bandwidth, dynamic range, accuracy, and the input/output voltage requirements. However, it should be noted at this point that high performance current mirrors must possess low input impedance, high output impedance and high accuracy under low-voltage operation. Since modern applications require very low input/output and supply voltages, most already published structures sustain performance degradation or even cannot operate in a low-voltage environment [6–8].

More specifically, the simple current mirror, shown in Fig. 1.1a, has very low input ($V_{in,min} = V_{th}$) and output ($V_{out,min} = V_{DS,sat}$) voltage requirements and needs at least a supply voltage $V_{DD,min} = V_{th} + V_{DS,sat}$. Even though this cell could be utilized in low-voltage systems, the output impedance of this current mirror is relatively low and is getting worse in scaled down technologies, making the use of this circuit impractical in most cases. The traditional way for increasing the output impedance of a current mirror is the use of cascode structures like the conventional cascode current mirror shown in Fig. 1.1b. But, the minimum input and output voltages of this cell are now $V_{in,min} = 2V_{th}$ and $V_{out,min} = 2V_{DS,sat}$, respectively, where the minimum supply voltage is $V_{DD,min} = 2V_{th} + V_{DS,sat}$. Therefore, the conventional cascode current mirror, as well as other similar topologies like the Wilson and improved Wilson current mirror, is not suitable for low-voltage operation. In addition, the reduction of supply voltage sets a constraint in the available input and output signal swings. The most common solution for achieving a sufficiently large output impedance in a current mirror without increasing $V_{DD,min}$ is the implementation shown in Fig. 1.1c [9–11]. Due to the cascode transistor in the output branch, this topology is characterized by very large output impedance, while due to the utilization of the Flipped Voltage Follower (FVF) at the input branch, requires the same minimum input voltage (i.e. $V_{in,min} = V_{th}$) as the simple current mirror. This structure is commonly denoted as low-voltage high-swing cascode current mirror and its minimum supply voltage is $V_{DD,min} = V_{th} + V_{DS,sat}$. Another attractive feature is the high copying accuracy since the drain currents of cascode transistors M_{n3}, M_{n4} are approximately equal, resulting in equal drain-source voltages of transistors M_{n1}, M_{n2} i.e. $V_{DS1} = V_{DS2}$. On the other hand, a drawback of this topology is the restriction imposed at the input signal from the dc bias voltage V_{DC} in order M_{n3}, M_{n4} to remain in saturation region. Moreover, even though the input signal swing is large, it is not the maximum due to the fact that the swing of M_{n1} is "strangled" by the V_{GS2} of the current sensing transistor M_{n2}, i.e. $V_{in,pp} = V_{th} - V_{DS,sat} = 2V_{th} - V_{GS,Mn2}$.

According to the above discussion, this specific structure of current mirror combines the low input voltage requirement and the high output signal swing, two key features in the design of low-voltage low-power analog circuits. Thus, this book focuses on the design of novel low-voltage analog integrated filters, where the basic element is the low-voltage high-swing current mirror. The obtained results from the proposed filter structures highlight another aspect of analog circuit design using current mirrors, where the majority of the existing publications are referred to conventional current mirrors [12–17]. Apart from the feature of low-voltage operation, attractive characteristics of the proposed topologies are their very simple structure and easy implementation of operations such as addition, inversion and amplification/attenuation of signals. Moreover, the resonant frequency of the proposed filters can be electronically controlled by modifying appropriate dc currents. Thus, any effect from process, voltage and temperature (PVT) variations can be eliminated. Finally, an advantage from the integration point of view is the derivation of resistorless filters due to the utilization of current mirrors' input resistance.

1.3   Organization of the Book

This book describes the design of low-voltage analog integrated filters using current mirrors that are applicable in modern low-power wireless and biomedical applications. The remainder of this book is organized as follows. Chap. 2 presents the structures of lossy and lossless integrators using current mirrors. Also, several topologies of second-order filters are introduced which are consisted by a small number of active elements and provide all the five standard filter functions namely lowpass, highpass, bandpass, bandstop and allpass from the same configuration. Some of the derived topologies offer the feature of orthogonal adjustment between the resonant frequency and the quality factor. Among the proposed second-order filters, those with the optimum performance have been fabricated in the AMS 0.35 μm n-well CMOS process.

Chap. 3 introduces the design of low-voltage complex filters for wireless receivers. The complex signal processing is used primarily in low Intermediate

Frequency (IF) receivers for removing the undesired image signals, resulted from the complex down conversion in the mixing of a Radio Frequency (RF) signal with the signal of Local Oscillator (LO). The implementation of complex filters is achieved by appropriately modifying the corresponding real filters. The derivation of novel complex filter topologies according to the Leapfrog and topological emulation techniques is thoroughly discussed where the employed active elements are low-voltage current mirrors. Also, according to the aforementioned filter design methods, two twelfth-order complex filters which satisfy the requirements of the Bluetooth standard are fabricated in the AMS 0.35 μm n-well CMOS process and are experimentally verified.

In Chap. 4, novel topologies suitable for realizing wavelet filter functions are studied. Wavelet filters are extensively used in biomedical applications and particularly in low-voltage low-power implantable devices for ECG signal processing. Due to the fact that they are constructed by current mirrors using MOS biased at subthreshold region, the proposed topologies offer the benefit of ultra-low voltage (0.5 V) operation, while the bias currents in this mode of operation are very low leading to reduced total power consumption. Two topologies of wavelet filters are presented. The first one is implemented by using the Follow the Leader Feedback (FLF) method while the second one using the operational emulation of an orthonormal filter. The efficiency of the proposed filters is verified through simulations results by employing TSMC 130 nm CMOS process.

In Chap. 5, a novel scheme for an adjustable CMOS current mirror is introduced. The proposed current mirror provides continuous gain adjustment, while it simultaneously features the attractive characteristic of low-voltage operation. It can be utilized in numerous applications such as the signal scaling and the design of electronically adjustable high-order filters. The behavior of the proposed topology has been experimentally verified through a first-order lowpass filter fabricated in the AMS 0.35 μm n-well CMOS process.

Finally, Chap. 6 presents the conclusions and some topics for further research in the design of analog circuits using low-voltage current mirrors.

References

1. "The International Technology Roadmap for Semiconductors (2009 edition)", ITRS, 2009 http://public.itrs.net.
2. E. Vittoz, "Future of analog in the VLSI environment", in *Proc. IEEE International Symposium on Circuits and Systems (ISCAS)*, pp. 1372–1375, May 1990.
3. K. Bult, "Analog design in deep sub-micron CMOS", in *Proc. European Solid-State Circuits Conference (ESSCIRC)*, pp.11–17, Sept. 2000.
4. Q. Huang, "Low voltage and low power aspects of data converter design", in *Proc. European Solid-State Circuits Conference (ESSCIRC)*, pp. 29–35, Sept. 2004.
5. C. Toumazou, F. Lidgey, and D.G. Haigh, "Analog IC design: The current-node approach", Peter Peregrenus Ltd., London, UK, 1990.

6. F. Yuan, and B. Sun, "A comparative study of low-voltage CMOS current-mode circuits for optical communications", in *Proc. Midwest Symposium on Circuits and Systems* (MWSCAS), vol. 1, pp. 315–319, Aug. 2002.

7. F. Ledesma, R. Garcia, and J. Ramirez-Angulo, "Comparison of new and conventional low-voltage current mirrors", in *Proc. Midwest Symposium on Circuits and Systems* (MWSCAS), vol. 2, pp. 49–52, Aug. 2002.

8. J. Ramirez-Angulo, R.G. Carvajal, and A. Torralba, "Low supply voltage high-performance CMOS current mirror with low input and output voltage requirements", *IEEE Transactions on Circuits and Systems II*, vol. 51, no. 3, pp. 124–129, Mar. 2004.

9. E. Sackinger, and W. Guggenbuhl, "A high-swing high-impedance MOS cascode circuit", *IEEE Journal of Solid-State Circuits*, vol. 25, no. 1, pp. 289–298, Feb. 1990.

10. R.G. Carvajal, J. Ramirez-Angulo, A. Torralba, J.A.G. Galan, A. Carlosena and F.M. Chavero, "The Flipped Voltage Follower: A useful Cell for Low-Voltage Low-Power Circuit Design", *IEEE Transactions on Circuits and Systems I*, vol. 52, no. 7, pp. 1276–1291, Jul. 2005.

11. C. Laoudias, C. Psychalinos, "Low-Voltage CMOS Current-Mode Filters Using Current Mirrors: Two Alternative Approaches", in *Proc. of 14th IEEE Mediterranean Electrotechnical Conference* (MELECON), Ajaccio (France), May 2008, pp. 435–440.

12. J. Ramirez-Angulo, M. Robinson, and E. Sanchez-Sinencio, "Current-mode continuous-time filters: two design approaches", *IEEE Transactions on Circuits and Systems II*, vol. 39, no. 6, pp. 337–341, Jun. 1992.

13. S.S. Lee, R.H. Zele, D.J. Allstot, and G. Liang, "CMOS Continuous-Time current-mode filters for high frequency applications", *IEEE Journal of Solid-State Circuits*, vol. 28, no. 3, pp. 323–329, Mar.1993.

14. A.H.M. Shousha, "Implementations of continuous-time current-mode ladder filters using multiple output current integrators", *International Journal of Electronics*, vol. 85, no. 4, pp. 497–509, 1998.

15. G. Souliotis, A. Chrisanthopoulos, and I. Haritantis, "Current Differential Mirrors: New circuits and applications", *International Journal of Circuit Theory Applications*, vol. 29, no. 6, pp. 553–574, Nov./Dec. 2001.

16. G. Souliotis, and C. Psychalinos, "Harmonic oscillators realized using current amplifiers and grounded capacitors", *International Journal of Circuit Theory and Applications*, vol. 35, no. 2, pp. 165–173, Mar. 2007.

17. G. Souliotis, and I. Haritantis, "Current-mode filters based on current mirror arrays", *International Journal of Circuit Theory and Applications*, vol. 36, no. 2, pp. 173–183, Mar. 2008.

Chapter 2
Universal Biquads Using Current Mirrors

Abstract This chapter presents a series of novel universal biquad topologies, where low-voltage current mirrors are employed as active elements. Due to current-mode nature of the performed signal processing, operations such as addition, subtraction, scaling and integration are very easily realized. Thus, the derived topologies consist of a small number of active elements contributing in lower power consumption, while they present the feature of low-voltage operation. The proposed universal biquads, which can be classified as either Single Input Multiple Output (SIMO) or Multiple Input Single Output (MISO), provide the five standard filter functions (lowpass, highpass, bandpass, bandstop and allpass) from the same configuration. Other advantages of the derived topologies are the absence of passive resistors, the resonant frequency of the filters can be electronically controlled by an appropriate dc current and only grounded capacitors are needed for the integrators. In addition, some of the proposed filters offer the feature of orthogonal adjustment between the resonant frequency and the quality factor. The behavior evaluation of the proposed universal biquads has been performed through a test chip prototype fabricated in AMS 0.35 μm CMOS technology.

Keywords Analog filters • Biquads • CMOS analog integrated circuits • Current mirror filters • Current mode circuits

2.1 Introduction

Current mirrors are the most simple and reliable building block for performing current-mode signal processing. The key performance feature of current mirrors is their inherent wide bandwidth capability since they do not present any high impedance internal node. In addition, the input resistance of current mirrors is used for realizing the required passive resistors leading also in the capability for the electronic adjustment of their values. Thus, an electronic tuning is possible in the case of analog filters because the realized time constants are controlled by a dc bias current [1–8].

C. Laoudias and C. Psychalinos, *Integrated Filters for Short Range Wireless and Biomedical Applications*, SpringerBriefs in Electrical and Computer Engineering, DOI 10.1007/978-1-4614-0260-2_2, © Springer Science+Business Media, LLC 2012

Biquad filters are very useful blocks to realize high-order filters by cascading multiple first and second-order sections. The second-order filter function, in its general form, is the following:

$$F(s) = \frac{a_2 s^2 + a_1 s + a_0}{s^2 + b_1 s + b_0} = K\frac{(s - z_1)(s - z_2)}{(s - p_1)(s - p_2)} \tag{2.1}$$

where z_i and p_i ($i = 1,2$) are the zeros and poles of the transfer function, respectively. F(s) can be written in an alternative biquadratic expression as

$$F(s) = K\frac{s^2 + \frac{\omega_{oz}}{Q_z} s + \omega_{oz}^2}{s^2 + \frac{\omega_{op}}{Q_p} s + \omega_{op}^2} \tag{2.2}$$

where ω_{op}, ω_{oz} are the undamped natural frequencies of the poles and zeros, respectively, while Q_p, Q_z are the corresponding quality factors, or Q factors. It is common to use ω_o and Q instead of ω_{op} and Q_p, so hereinafter this notation will be used. The pole or zero frequency is the magnitude of the pole or zero, respectively. A key role in the behavior of second-order filters has the Q factor which determines how near the $j\omega$-axis is the corresponding pole in the s-plane, and whether the poles will be complex conjugate or real. The poles of (2.2) are

$$s_{1,2} = \omega_o\left(-\frac{1}{2Q} \pm \sqrt{\left(\frac{1}{2Q}\right)^2 - 1}\right). \tag{2.3}$$

It is apparent from (2.3) that the realization of active second-order filters is of interest only in the case that $Q > 0.5$, i.e. when the poles are complex conjugate. Otherwise, when the poles are negative real, the realization can be achieved by using passive RC networks only. Depending on the positions of zeros in the s-plane, the second-order filter frequency responses of interest obtained from (2.2) are lowpass (LP), highpass (HP), bandpass (BP), bandstop (BS) and allpass (AP).

Universal biquads are defined as second-order filters where the five standard transfer functions are provided by the same topology. They are very useful building blocks in analog signal processing and particularly in the fields of electronic measurement, communication, auto-control and nerve networks. More specifically, they could be used in the implementation of touchtone telephone decoders, phase-locked loops, FM stereo demodulators and crossover networks [9, 10]. For example, a crossover network used in a three-way high fidelity loudspeaker consists of a lowpass, a bandpass and a highpass filter as shown in Fig. 2.1. Thus, each speaker receives frequencies in the range in which it is most efficient. Most loudspeakers use multiple drivers (tweeter for high frequencies, woofer for low frequencies and mid-range for intermediate frequencies) and employ crossover networks to route the appropriate frequency band to the corresponding driver.

Fig. 2.1 Three-way high
fidelity loudspeaker crossover
network

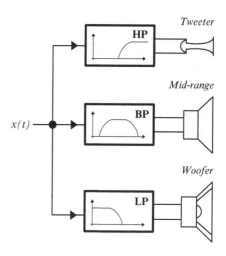

Over the last few years, current amplifiers of low gain have been used under various names, such as current followers, unity gain cells, current buffers etc., for the realization of transfer function for filtering applications. A significant research effort has been carried out in order to realize universal biquads with active elements configured as current mirrors. In [11], a current-mode version of the Tow-Tomas biquad filter has been realized by employing CCIIs, as active elements, where their Y terminal is grounded. It is obvious from this configuration that the employed CCIIs behave as current mirrors, while passive resistors are needed to obtain the required integrators. The employment of CCIIs configured as current followers has also been performed in [12, 13]. A drawback of these topologies is the requirement of floating capacitors leading to a limited frequency operation caused by the parasitic capacitances effect. Another drawback of these topologies is that they cannot be considered as universal biquads due to the fact that all the five standard transfer functions are not simultaneously offered. In addition, a manipulation of their structures is needed in order to derive the AP transfer function. Actual universal biquads could also be realized by simultaneously employing both types of unity-gain cells i.e. current followers and voltage followers [14–19]. Finally, several voltage-mode universal biquads have been proposed using CFOAs [20] and OTAs [21, 22] as filter building blocks.

Conventional current mirrors have been already employed for designing MISO biquad filter topologies [2, 23]. The main drawback of these topologies is that manipulation of their core structure must be performed in order to become an actual universal biquad. More specifically, an additional current inverter must be utilized for the derivation of the AP transfer function, making these topologies not feasible in practice.

The contribution made by this work is that novel MISO and SIMO universal biquads using low-voltage current mirrors are designed. The proposed MISO universal biquads, compared with the already published topologies in [2, 23], do not require any modification of their structures, while they present improved performance characteristics in terms of total power dissipation, noise, dynamic range and sensitivity in active and passive mismatching. Moreover, it is worth mentioning that SIMO universal biquads using current mirrors have never been reported in the literature. As in [2, 23], passive resistors are avoided due to the fact that the input resistance of current mirrors is employed, whereas the resonant frequency can be electronically adjusted, which is originated from the small-signal nature of the input resistance. Other benefits are the capability of orthogonal adjustment between the resonant frequency and the quality factor of the derived filters and the potential for low-voltage operation. The behavior of the proposed filters is evaluated through simulation results, where the most important performance factors have been considered. The two topologies with the most optimum performance have been fabricated in AMS 0.35 μm C35B4C3 CMOS process and verified through extensive experimental measurements.

2.2   Proposed Universal Biquad Topologies

This section shows the implementation of universal biquad filters that are realized by a two-integrator loop. Based on this technique, the resonant frequency of the derived filters can be adjusted without disturbing the quality factor, while the opposite does not hold. However, the use of current mirrors allows in some of the proposed universal biquads the independent control of these two parameters i.e. the orthogonal adjustment between ω_0 and Q.

Thus, the corresponding biquad transfer functions can be formed by the appropriate configuration of multiple output lossy and lossless integrators. As it has already mentioned, a SIMO universal biquad provides all the five standard filter function simultaneously at a different node of its topology, while a MISO topology provides a different transfer function by altering the way that input signals are connected.

2.2.1   Integrator Blocks Using Current Mirrors

A realization of multiple output lossy and lossless integrators using current mirrors as active elements is depicted in Fig. 2.2a, b, respectively, while their corresponding symbols are given in Fig. 2.3. The required integrators' time constants are formed by the grounded capacitor and the input resistance of current mirror. Note that the latter is equal to $1/g_m$, where g_m is the transconductance parameter of an MOS operating in saturation region. The small-signal transconductance parameter is given by the formula $g_m = \sqrt{2K\frac{W}{L}I_0}$, where $K = \mu_n C_{ox}$ is the transconductance factor of the MOS transistor, μ_n is the low-field electron mobility and C_{ox} is the oxide capacitance per unit channel area. Thus, the realized

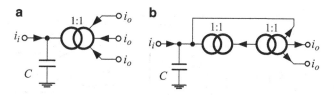

Fig. 2.2 Realization of multiple output (**a**) lossy and (**b**) lossless integrator using current mirrors

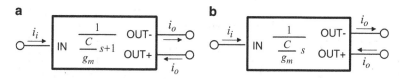

Fig. 2.3 Symbol of multiple output (**a**) lossy and (**b**) lossless integrator using current mirrors

time constant (τ) in both configurations is equal to C/g_m, where it is obvious that it can be controlled by the bias current I_0 of current mirrors.

2.2.2   SIMO Universal Biquads

Using the symbols of the integrators from Fig. 2.3, the Functional Block Diagrams (FBD) of the proposed SIMO universal biquads are given in Fig. 2.4 [24–27].

Each topology is comprised by a multiple output current mirror, denoted as CM, in order to perform the required current distribution and summation of the input current, a lossy and a lossless integrator. Due to the employment of current mirrors passive resistors are avoided, while only grounded capacitors are required. By performing a routine algebraic analysis of FBDs in Fig. 2.4a, b, the following transfer functions are derived

$$H_{LP}(s) = \frac{\frac{g_{m1}g_{m2}}{C_1 C_2}}{s^2 + \frac{g_{m1}}{C_1} s + \frac{g_{m1}g_{m2}}{C_1 C_2}} \qquad (2.4a)$$

$$H_{HP}(s) = \frac{s^2}{s^2 + \frac{g_{m1}}{C_1} s + \frac{g_{m1}g_{m2}}{C_1 C_2}} \qquad (2.4b)$$

$$H_{BP}(s) = \frac{\frac{g_{m1}}{C_1} s}{s^2 + \frac{g_{m1}}{C_1} s + \frac{g_{m1}g_{m2}}{C_1 C_2}} \qquad (2.4c)$$

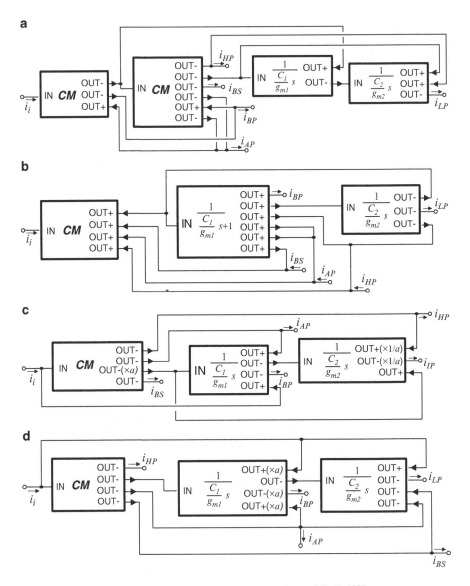

Fig. 2.4 Proposed SIMO universal biquads (**a**) [24] (**b**) [25] (**c**) [26] (**d**) [27]

$$H_{BS}(s) = \frac{s^2 + \frac{g_{m1}g_{m2}}{C_1C_2}}{s^2 + \frac{g_{m1}}{C_1}s + \frac{g_{m1}g_{m2}}{C_1C_2}} \tag{2.4d}$$

$$H_{AP}(s) = \frac{s^2 - \frac{g_{m1}}{C_1}s + \frac{g_{m1}g_{m2}}{C_1C_2}}{s^2 + \frac{g_{m1}}{C_1}s + \frac{g_{m1}g_{m2}}{C_1C_2}} \tag{2.4e}$$

where variables g_{m1} and g_{m2} are the transconductance parameters of the corresponding input transistor of each integrator. The expressions of ω_0 and Q are expressed by (2.5) and (2.6), respectively.

$$\omega_o = \sqrt{\frac{g_{m1}g_{m2}}{C_1 C_2}} \tag{2.5}$$

$$Q = \sqrt{\frac{g_{m2}C_1}{g_{m1}C_2}} \tag{2.6}$$

From (2.5) and (2.6), the following remarks can be made: (1) the parameter ω_0 can be electronically adjusted through the dc bias current I_0 of current mirrors and (2) assuming that g_{m1} and g_{m2}, the parameter ω_0 can be adjusted without disturbing the parameter Q. Note that in the second case the Q factor of the filters is realized by a capacitor ratio and, consequently, careful layout techniques (e.g. common centroid) must be performed in order to achieve an appropriate level of accuracy.

The realized LP, HP, BP, BS and AP transfer functions provided by the FBDs in Fig. 2.4c, d are given by

$$H_{LP}(s) = \frac{\frac{g_{m1}g_{m2}}{C_1 C_2}}{s^2 + \alpha \cdot \frac{g_{m1}}{C_1}s + \frac{g_{m1}g_{m2}}{C_1 C_2}} \tag{2.7a}$$

$$H_{HP}(s) = \frac{s^2}{s^2 + \alpha \cdot \frac{g_{m1}}{C_1}s + \frac{g_{m1}g_{m2}}{C_1 C_2}} \tag{2.7b}$$

$$H_{BP}(s) = \frac{\alpha \cdot \frac{g_{m1}}{C_1}s}{s^2 + \alpha \cdot \frac{g_{m1}}{C_1}s + \frac{g_{m1}g_{m2}}{C_1 C_2}} \tag{2.7c}$$

$$H_{BS}(s) = \frac{s^2 + \frac{g_{m1}g_{m2}}{C_1 C_2}}{s^2 + \alpha \cdot \frac{g_{m1}}{C_1}s + \frac{g_{m1}g_{m2}}{C_1 C_2}} \tag{2.7d}$$

$$H_{AP}(s) = \frac{s^2 - \alpha \cdot \frac{g_{m1}}{C_1}s + \frac{g_{m1}g_{m2}}{C_1 C_2}}{s^2 + \alpha \cdot \frac{g_{m1}}{C_1}s + \frac{g_{m1}g_{m2}}{C_1 C_2}} \tag{2.7e}$$

where variables g_{m1} and g_{m2} have the same meaning as in (2.4a)–(2.4e). The current scaling factor α is implemented by an appropriate configuration of the corresponding outputs of current mirrors, i.e. the aspect ratio and dc bias current of transistors must be accordingly scaled. Consequently, a careful layout (e.g. usage of arrays of unit transistors) must be performed in order to realize high accuracy in

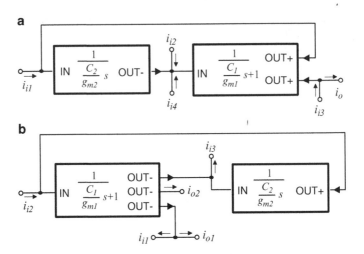

Fig. 2.5 Proposed MISO universal biquads (**a**) [25] (**b**) [28]

transistor aspect ratios. The expression of the resonant frequency is still given by (2.5), while the quality factor is expressed by (2.8) as

$$Q = \frac{1}{a}\sqrt{\frac{g_{m2}C_1}{g_{m1}C_2}} \qquad (2.8)$$

From (2.5) and (2.8), the following remarks can be made: (1) the parameter ω_0 can be electronically adjusted through the dc bias current I_0 of current mirrors (2) the parameter ω_0 can be adjusted without disturbing the Q factor of the filters, due to the fact that the ratio of transconductances is independent of I_0 and (3) the Q factor could be adjusted through the factor α without disturbing ω_0. In other words, the advantageous characteristic offered by the topologies in Fig. 2.4c, d is that parameters ω_0 and Q are orthogonal without any restriction imposed.

2.2.3 MISO Universal Biquads

In a similar way, the derived topologies of MISO universal biquads are presented in Fig. 2.5 [25, 28]. The expression of the output current i_o in Fig 2.5a is

$$i_o(s) = \frac{\left(s^2 + \frac{g_{m1}}{C_1}s + \frac{g_{m1}g_{m2}}{C_1C_2}\right)i_{i3} - \frac{g_{m1}}{C_1}s\cdot\left(i_{i2} + i_{i4}\right) - \frac{g_{m1}g_{m2}}{C_1C_2}i_{i1}}{s^2 + \frac{g_{m1}}{C_1}s + \frac{g_{m1}g_{m2}}{C_1C_2}} \qquad (2.9)$$

where variables g_{m1} and g_{m2} have the same meaning as in previous. The resonant frequency and quality factor are given by (2.5) and (2.6), respectively. As a result, the same conclusions hold about the orthogonal dependency between parameters ω_0 and Q and electronic tunability of ω_0 as in topologies of Fig. 2.4a, b. From (2.9), it is revealed that in order to achieve the five standard filter functions, the following conditions must be satisfied:

- Lowpass filter: $i_{i2} = i_{i3} = i_{i4} = 0$, $i_{i1} = i_i$
- Highpass filter: $i_{i4} = 0$, $i_{i1} = i_{i2} = i_{i3} = i_i$
- Bandpass filter: $i_{i1} = i_{i3} = i_{i4} = 0$, $i_{i2} = i_i$
- Bandstop filter: $i_{i1} = i_{i4} = 0$, $i_{i2} = i_{i3} = i_i$
- Allpass filter: $i_{i1} = 0$, $i_{i2} = i_{i3} = i_{i4} = i_i$

It should be mentioned that, if the input current i_{o2} would be equal to $2i_i$, then the input current i_4 could be omitted. In other words, the topology could be considered as a three-input MISO universal biquad.

The expressions of the output currents i_{o1} and i_{o2} in Fig. 2.5b are

$$i_{o1}(s) = -\frac{\left(s^2 + \frac{g_{m1}}{C_1}s + \frac{g_{m1}g_{m2}}{C_1C_2}\right)i_{i1} - \frac{g_{m1}}{C_1}s \cdot i_{i2} - \frac{g_{m1}g_{m2}}{C_1C_2}i_{i3}}{s^2 + \frac{g_{m1}}{C_1}s + \frac{g_{m1}g_{m2}}{C_1C_2}} \tag{2.10a}$$

$$i_{o2}(s) = \frac{\frac{g_{m1}}{C_1}s \cdot i_{i2} + \frac{g_{m1}g_{m2}}{C_1C_2}i_{i3}}{s^2 + \frac{g_{m1}}{C_1}s + \frac{g_{m1}g_{m2}}{C_1C_2}} \tag{2.10b}$$

where the same remarks hold about parameters ω_0 and Q as in topologies of Figs. 2.4a, b and 2.5a. In (2.10a)–(2.10b), the five filter functions can be obtained as

- Lowpass filter: $i_{in1} = i_{in2} = 0$, $i_{in3} = i_{in}$, $i_o = i_o = i_{o2}$
- Highpass filter: $i_{in1} = i_{in2} = i_{in3} = i_{in}$, $i_o = i_{o1}$
- Bandpass filter: $i_{in1} = i_{in3} = 0$, $i_{in2} = i_{in}$, $i_o = i_{o1} = i_{o2}$
- Bandstop filter: $i_{in1} = i_{in2} = i_{in}$, $i_{in3} = 0$, $i_o = i_{o1}$
- Allpass filter: $i_{in1} = i_{in2} = i_{in}$, $i_{in3} = 0$, $i_o = i_{o1} + i_{o2}$

It must be noted that the AP filter function can be easily obtained by establishing an appropriate simple short circuit between the outputs that provide the currents i_{o1} and i_{o2}. In addition, by considering the current i_{o1} as the output signal, the topology can be viewed as a MISO universal biquad. Finally, it is obvious that in the case of MISO topologies in Fig. 2.5c, d, where a different transfer function is provided each time by altering the way that input signals are connected, a practical approach for the selection of the corresponding transfer function is the utilization of a digitally controlled CMOS switch pattern. Thus, depending on the state of the switch (ON/OFF), the appropriate input current will be provided each time.

2.3 Comparison Results

In order to simulate the operation of the proposed topologies, the low-voltage high-swing cascode current mirror as that given in Fig. 2.6 [29] will be employed. The input conductance g_i of the current mirror is equal to the sum of the transconductance of transistor M_{n11} and output conductance of transistor M_{p1}

$$g_i = g_{m,M_{n11}} + g_{ds,M_{p1}} \tag{2.11}$$

The output conductance g_o of the current mirror is given by the following expression

$$g_o = \left(g_{ds,M_{n12}} \cdot g_{ds,M_{n22}} \big/ g_{m,M_{n22}}\right) + g_{ds,M_{p2}} \tag{2.12}$$

As mentioned in the previous chapter, the minimum supply voltage of this cell is $V_{DD,min} = V_{th} + V_{DS,sat}$. Thus, the current mirrors are biased at a single power supply voltage $V_{DD} = 1.5$ V, while the dc voltage V_{DC} is chosen to be 1.2 V. Considering a dc bias current equal to $I_0 = 4$ μA, the corresponding values of NMOS transistors aspect ratios are 8 μm/1 μm for M_{n1i} and 7 μm/0.6 μm for M_{n2i} ($i = 1,2,\ldots$). The PMOS transistors M_{pi} ($i = 1,2,\ldots$) have an aspect ratio 8 μm/1 μm. Under the above conditions, the achieved input resistance of current mirrors is 14.32 kΩ. In order to realize a filter with a resonant frequency 1 MHz and a Q factor equal to 2, the capacitor values are chosen to be as: $C_1 = 22.24$ pF and $C_2 = 5.56$ pF.

The operation of the proposed topologies has been assessed through simulation results in Cadence Analog Design Environment using Spectre simulator. For this purpose, the AMS 0.35 μm CMOS process has been employed, where the most important factors including linearity, noise, mismatching etc. will be evaluated. It must be noted that, the MISO topology in [2] which is realized by employing the

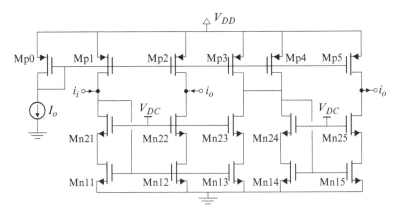

Fig. 2.6 Low-voltage multiple-output current mirror topology

same configuration of low voltage current mirror and the topology derived according to the FBD in Fig. 2.4d by employing the MOS version of the CCCII introduced in [30] have been also considered. In order to achieve fair comparison results, the same bias scheme about voltages and currents has been utilized for both realizations, whereas the transistor sizing has been kept the same as in the proposed universal biquads. Owing to the fact that the biquad realized using the CCCII in [30] does not have the capability to operate in a single 1.5 V power supply voltage, a symmetrical ±1.5 V[1] bias scheme has been utilized.

The linearity of the realized universal biquads, concerning the bandpass filter response, is evaluated through the third-order intermodulation distortion (IM3). Two closely spaced tones with variable amplitude and frequencies 1 MHz and 1.005 MHz have been applied at their inputs in order to determine the input power P_{in} where IM3 is equal to 1% (-40 dB). The third-order Input Intercept Point (IIP3) is defined as

$$IIP3 = P_{in} + \frac{IM3}{2} \tag{2.13}$$

The input referred noise (INOISE) integrated over a 1 MHz range has been calculated for each topology. Thus, the SFDR is given by the following expression

$$SFDR = (2/3) \cdot (IIP3 - INOISE) \tag{2.14}$$

Moreover, Monte Carlo analysis has been carried out in order to verify the sensitivities of the filters in active and passive mismatching. Thus, by performing 100 iterations and taking into account both mismatch and process variations, the standard deviations of the following three characteristics have been resulted: (1) maximum gain at center frequency (2) center frequency and (3) bandwidth. The obtained results are summarized in Tables 2.1 and 2.2.

Considering the results in Tables 2.1 and 2.2, the following conclusions can be made: (1) the universal biquads that are constructed by current mirrors have the same IIP3, (2) the SIMO universal biquad in Fig. 2.4b has better performance in all the terms under consideration compared with that achieved by the other topologies, (3) the MISO universal biquad in Fig. 2.5a has the best performance in all the terms under consideration and more specifically, it offers an almost 50% reduction of the required power dissipation with that required in [2] and (4) the only drawback in the topologies of Figs. 2.4a, b and 2.5a, b is that the orthogonal adjustment of ω_0 and Q is achieved under the condition of equal transconductance parameters g_{m1} and g_{m2}.

Thus, it can be concluded that the SIMO universal biquad of Fig. 2.4b and the MISO universal biquad of Fig. 2.5a are the two proposed filters with the most optimum performance, while their topologies derived using the configurations of Fig. 2.2 are given in Fig. 2.7. The effects of the current mirror non-idealities in the transfer functions of these two topologies are studied in detail in the next section.

---

[1] The minimum supply voltage of CCCII is equal to $V_{DD,min} = 2V_{th} + 2V_{DS,sat}$.

Table 2.1 Comparison results for the universal SIMO biquads

Performance factors	Fig. 2.4a	Fig. 2.4b	Fig. 2.4c	Fig. 2.4d	[30]
Power dissipation (μW)	182.2	124	142	148	350
Total capacitance (pF)	27.8	27.8	27.8	27.8	66
Input referred noise (nA)	8.35	6.73	8.12	8.8	9.25
IIP3 (dBm)	−96.4	−96.6	−96.6	−97.1	−89.2
SFDR (dB)	34.7	35.9	34.8	34	39
Sensitivity of ω_0 (kHz)	118.5	88.2	135.43	82.5	57.4
Sensitivity of gain	0.4	0.3	0.3	0.6	0.1
Sensitivity of BW (kHz)	133.2	126.6	136.7	155.1	41.8
Orthogonal ω_0, Q	✔	✔	☑	☑	☑

✔: When $g_{m1} = g_{m2}$; ☑: Unconditional

Table 2.2 Comparison results for MISO biquads

Performance factors	Fig. 2.5a	Fig. 2.5b	[2]
Power dissipation (μW)	61	88.5	106.6
Total capacitance (pF)	27.8	27.8	27.8
Input referred noise (nA)	6.49	8.11	7.79
IIP3 (dBm)	−96.8	−96.8	−96.7
SFDR (dB)	36	34.7	35
Sensitivity of ω_0 (kHz)	85.6	127.7	81.7
Sensitivity of gain	0.2	0.25	0.38
Sensitivity of BW (kHz)	125.4	122.9	146
Orthogonal ω_0, Q	✔	✔	☑

✔: When $g_{m1} = g_{m2}$; ☑: Unconditional

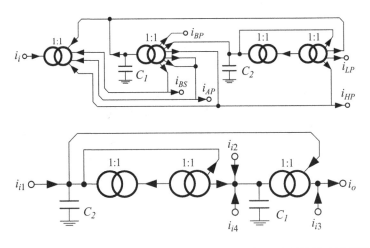

Fig. 2.7 Proposed universal biquad topologies using current mirrors (**a**) SIMO (**b**) MISO

2.3.1 Effects of Current Mirror Non-idealities

So far, the operation of universal biquads has been studied neglecting the transistor mismatches and second-order effects, like the body effect. In addition, the finite input–output conductance ratio of a current mirror degrades its performance in terms of current transfer accuracy. Therefore, the transfer function of a unity gain current mirror is given by

$$\frac{i_o}{i_i} = \frac{1}{1+\varepsilon} \cong 1 - \varepsilon \tag{2.15}$$

where $\varepsilon \equiv g_o/g_i$. Variables g_i and g_o are the input and output conductance of the current mirror,[2] respectively. Taking into account the effect of the input–output conductance ratio ε and reanalyzing both biquads in Fig. 2.7, the derived expressions that correspond to (2.4) and (2.9) are given by (2.16a)–(2.16e) and (2.17a), respectively.

$$\hat{H}_{LP}(s) \cong \frac{(1-4\varepsilon)\dfrac{g_{m1}g_{m2}}{C_1C_2}}{s^2 + \dfrac{\hat{\omega}_o}{\hat{Q}}s + \hat{\omega}_o^2} \tag{2.16a}$$

$$\hat{H}_{HP}(s) \cong -\frac{s^2 + 2\varepsilon\dfrac{g_{m1}}{C_1}s + \varepsilon\dfrac{g_{m1}g_{m2}}{C_1C_2}}{s^2 + \dfrac{\hat{\omega}_o}{\hat{Q}}s + \hat{\omega}_o^2} \tag{2.16b}$$

$$\hat{H}_{BP}(s) \cong \frac{(1-2\varepsilon)\left(\dfrac{g_{m1}}{C_1}\right)s + 2\varepsilon\dfrac{g_{m1}g_{m2}}{C_1C_2}}{s^2 + \dfrac{\hat{\omega}_o}{\hat{Q}}s + \hat{\omega}_o^2} \tag{2.16c}$$

$$\hat{H}_{BS}(s) \cong -\frac{s^2 + 2\varepsilon\left(\dfrac{g_{m1}}{C_1} + \dfrac{g_{m2}}{C_2}\right)s + (1-3\varepsilon)\dfrac{g_{m1}g_{m2}}{C_1C_2}}{s^2 + \dfrac{\hat{\omega}_o}{\hat{Q}}s + \hat{\omega}_o^2} \tag{2.16d}$$

$$\hat{H}_{AP}(s) \cong -\frac{s^2 - \left[(1-4\varepsilon)\dfrac{g_{m1}}{C_1} - 2\varepsilon\left(\dfrac{g_{m2}}{C_2}\right)\right]s + (1-5\varepsilon)\dfrac{g_{m1}g_{m2}}{C_1C_2}}{s^2 + \dfrac{\hat{\omega}_o}{\hat{Q}}s + \hat{\omega}_o^2} \tag{2.16e}$$

---

[2] See (2.11), (2.12).

$$\hat{i}_o(s) \cong \frac{\alpha' i_{i3} - \alpha'' \cdot (i_{i2} + i_{i4}) - \alpha''' i_{i1}}{s^2 + \dfrac{\hat{\omega}_o}{\hat{Q}} s + \hat{\omega}_o^2} \tag{2.17a}$$

where variables a', a'', and a''' are expressed as

$$\alpha' = s^2 + \left(\frac{g_{m1}}{C_1} + 2\varepsilon \frac{g_{m2}}{C_2}\right) s + (1 - \varepsilon) \frac{g_{m1} g_{m2}}{C_1 C_2} \tag{2.17b}$$

$$\alpha'' = \left[(1 - \varepsilon)\frac{g_{m1}}{C_1} + 2\varepsilon \frac{g_{m2}}{C_2}\right] s \tag{2.17c}$$

$$\alpha''' = (1 - 3\varepsilon) \frac{g_{m1} g_{m2}}{C_1 C_2} \tag{2.17d}$$

The expressions for the resonant frequency $\hat{\omega}_0$ and $\hat{Q}$ factor for both filters are now given by (2.18) and (2.19) as

$$\hat{\omega}_o = \omega_o \cdot \sqrt{1 - \varepsilon} \tag{2.18}$$

$$\hat{Q} = \sqrt{\frac{1 - \varepsilon}{4\varepsilon + \frac{1}{Q^2}}} \tag{2.19}$$

where ω_0 and Q are the nominal values of the resonant frequency and the quality factor expressed by (2.5) and (2.6), respectively.

From (2.16) to (2.17), it is obvious that the finite input–output conductance ratio of a current mirror significantly affects all the realized transfer functions. The resonant frequency deviates from its nominal value according to the expression in (2.18). With regards to the Q factor, the following points could be derived: for relative small values of Q (i.e. $Q^2 \ll 1/4\varepsilon$) the expression in (2.19) could be written in the form $\hat{Q} \cong Q\sqrt{1 - \varepsilon}$. Due to the fact that both ω_o and Q variables are affected from the same error factor, it can be concluded that the orthogonal adjustment between them is preserved. This is not the case for relative high Q values (i.e. $Q^2 \gg 1/4\varepsilon$) where from (2.19) is derived that $\hat{Q} \cong \sqrt{\frac{1-\varepsilon}{4\varepsilon}}$. According to this, the proposed topologies are not able to realize high Q filter functions due to the fact that the realized value of $\hat{Q}$ is only determined by the error factor, and not by the nominal value of Q.

The capability of both SIMO and MISO filters for electronic adjustment of the resonant frequency ω_0 without disturbing the quality factor Q is verified through simulation results. For this purpose the frequency responses of the bandpass filters tuned at 0.89, 1, 1.43 and 1.73 MHz are given in Fig. 2.8. The deviations of Q values with regards to the corresponding nominal values are summarized in Table 2.3. According to this Table it is obvious that capability of filters for orthogonal adjustment between ω_o and Q is degraded in the case of high Q filters. This is caused by the effect of finite input–output conductance ratio of current mirror, that imposes an error factor (ε) equal to $3.65 \cdot 10^{-3}$, under the employed biasing scheme. This factor could be further reduced in the case that cascoded PMOS biasing transistors would be employed for realizing the required dc current sources. It should be also noted at this point, that the approximation $\hat{Q} \cong Q\sqrt{1-\varepsilon}$ is valid for $Q \ll 8$.

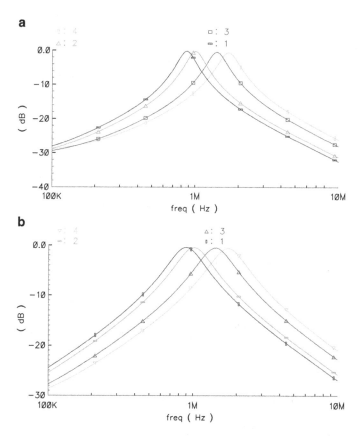

Fig. 2.8 Demonstration of orthogonal electronic adjustment of ω_0 and Q for SIMO (**a**) $Q = 2$ (**b**) $Q = 5$ and MISO (**c**) $Q = 2$ (**d**) $Q = 5$ universal biquad (**1**: $f_o = 0.89$ MHz, **2**: 1 MHz, **3**: 1.43 MHz and **4**: 1.73 MHz)

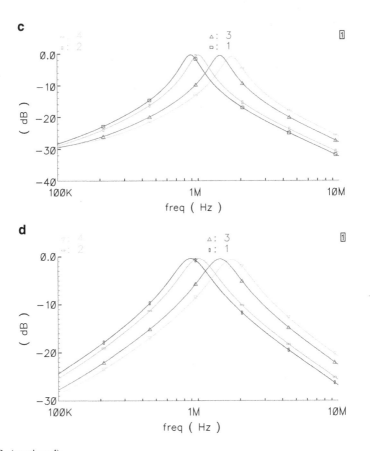

Fig. 2.8 (continued)

Table 2.3 Deviation of Q factor in SIMO an5d MISO filters

	SIMO				MISO			
	$\Delta Q/Q$				$\Delta Q/Q$			
Nominal value	$f_o = 0.89$ MHz	$f_o = 1$ MHz	$f_o = 1.43$ MHz	$f_o = 1.73$ MHz	$f_o = 0.89$ MHz	$f_o = 1$ MHz	$f_o = 1.43$ MHz	$f_o = 1.73$ MHz
$Q = 2$ (%)	3.85	5	5	7.5	2	5	5	8.5
$Q = 5$ (%)	11	20	24	26	11	22	26	28

2.4 Experimental Results

Differential filters are more preferable than the conventional single-ended filters in modern analog processing systems. This is originated from the fact that the differential structures offer improved performance in terms of linearity and noise rejection [31]. Thus, the proposed universal SIMO and MISO biquads in Fig. 2.7 were fabricated in differential form through the AMS C35B4C3 0.35 μm n-well CMOS process.

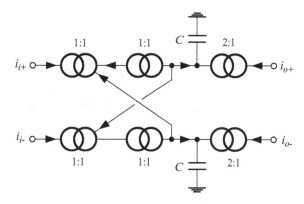

Fig. 2.9 Differential low-voltage lossy integrator using current mirrors

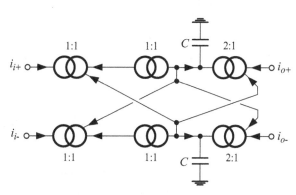

Fig. 2.10 Differential low-voltage lossless integrator using current mirrors

2.4.1 Differential Integrators Using Current Mirrors

A fully balanced lossy integrator consisting of a pair of cross-coupled current mirrors is shown in Fig. 2.9. The expression for the differential output current of the lossy integrator is given by (2.20) as,

$$i_{o+} - i_{o-} = \frac{1}{\tau s + 1}(i_{i+} - i_{i-}) \tag{2.20}$$

where the time constant τ is still given by the formula $\tau = C/g_m$.

In order to realize a differential lossless integrator, an enhanced circuitry compared to that in the structure of Fig. 2.9 is obtained, by employing an additional pair of cross-coupled current mirror. The derived topology is depicted in Fig. 2.10. Thus, the differential output current of the lossless integrator is expressed by (2.21),

$$i_{o+} - i_{o-} = \frac{1}{\tau s}(i_{i+} - i_{i-}) \tag{2.21}$$

where the time constant is given by the same formula as in the case of lossy integrator.

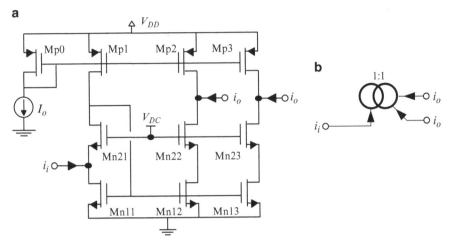

Fig. 2.11 (**a**) Conventional low-voltage cascode current mirror (**b**) the followed notation

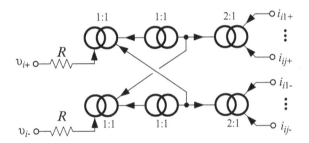

Fig. 2.12 Voltage-to-current conversion and current distribution in SIMO biquad

2.4.2 Voltage-to-Current Conversion in SIMO Biquad

Due to current-mode nature of the signal processing performed by the current mirrors, appropriate interfaces must be added at both input and output of the filters for the V-I and I-V conversion, respectively. These must fulfill the following requirements: (1) to be compatible with the integrators' bias scheme and current signal levels, (2) to contribute as small as possible in the distortion of the whole system and (3) to provide proper input and output impedance levels.

In the case of SIMO biquad topology, the differential input voltage-to-current conversion is realized via two off-chip resistors R. In order to achieve an accurate conversion and current distribution, the current is fed into a current mirror with low input impedance. Thus, the very simple structure of a conventional low-voltage cascode current mirror, known as Flipped Voltage Follower Current Sensor (FVFCS) is used as an input stage, where its topology and the followed symbol are shown in Fig. 2.11a, b. The input resistance of this current mirror is equal to $R_{in} = (g_{ds,Mn21} + g_{ds,Mp1})/(g_{m,Mn11} \cdot g_{m,Mn21})$.

Therefore, the operation of the first block of the SIMO universal biquad in Fig. 2.7a is realized by utilizing the scheme in Fig. 2.12.

Fig. 2.13 (a) Digitally
programmable current mirror,
(b) the followed notation

Fig. 2.14 Voltage-to-current
conversion and current
distribution in MISO
universal biquad

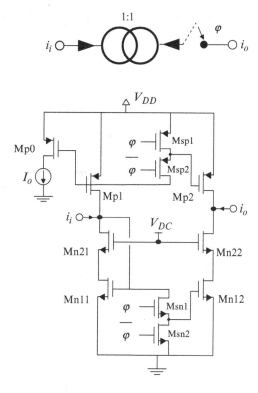

It is apparent that multiple current outputs are achieved by making an appropriate number of replicas where their differential expression is given by the formula $i_{ij+} - i_{ij-} = (v_{i+} - v_{i-})/R$, $(j = 1,\ldots,n)$. Moreover, due to the fact that the dc voltage at the current mirror input is sensitive to process and temperature variation, dc-decoupling capacitors are added in series with the off-chip resistors R.

2.4.3 Voltage-to-Current Conversion in MISO Biquad

In the case of MISO biquad topology, a similar approach as in SIMO biquad holds, concerning the differential input voltage-to-current conversion and feeding of the current signal. However, the only difference is related with current distribution due to the multiple input nature of MISO biquad. For this purpose, a digitally programmable differential current mirror is employed, in order to generate the four input currents i_{ij}, $(j = 1,\ldots,4)$. Its structure and the corresponding notation are shown in Fig. 2.13a, b, respectively. Thus, each input current i_{ij} of the four-input MISO topology is selected according to the status of φ_j, $(j = 1,\ldots,4)$ of the corresponding CMOS switch; it should be noted that $\bar{\varphi}_j$ is the complement of φ_j. In addition, when φ_j is in high-level, the input current i_{ij} is provided. The derived scheme of voltage-to-current conversion and current distribution using the digitally programmable current mirror is shown in Fig. 2.14. The aspect ratios of PMOS and NMOS transistors M_{sp} and M_{sn} were 10 μm/0.35 μm.

Table 2.4 Switch status condition for MISO biquad

Control bits $b_2 b_1 b_0$	Φ_1	Φ_2	Φ_3	Φ_4	Filter function
000	1	0	0	0	LP
001	1	1	1	0	HP
010	0	1	1	0	BS
011	1	0	1	1	BP
100	0	1	1	1	AP

1: High-level; 0: Low-level

Fig. 2.15 Circuit scheme of filter function selection in MISO universal biquad (DLC1)

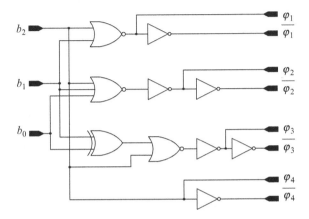

According to (2.10) and the conditions concerning the input currents given in Sect. 2.3, three digital control bits b_i ($i = 0,1,2$) are required to select one of the five filter transfer functions. Therefore, the switch status condition provided in Table 2.4 has to be followed in order to derive the appropriate filter function. One possible implementation of the required switching scheme is shown in Fig. 2.15 (*DLC1*).

2.4.4 Current-to-Voltage Conversion in SIMO and MISO Biquads

In order to maintain the fully balanced nature of the proposed biquad topologies, the differential output current must be converted into a differential voltage. For this purpose, an on-chip high-bandwidth two-stage transresistance amplifier as that depicted in Fig. 2.16 has been employed. It is biased at a single power supply voltage $V_{DD} = 1.5$ V, while the common-mode dc voltage (V_{cm}) was chosen to be 1 V. Considering a dc bias current equal to $I_o = 10$ μA, the corresponding values of the NMOS and PMOS transistor aspect ratios as well as the other components values are given in Table 2.5.

At this point, it should be noted that due to limited die area the implementation of only one transresistance amplifier is possible. Therefore, a similar approach to

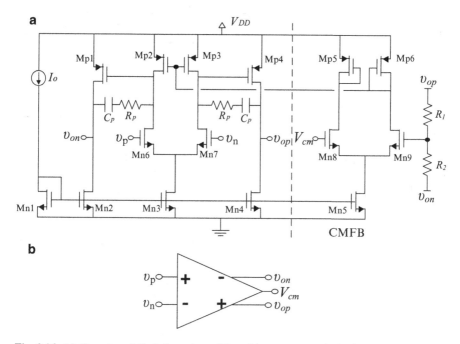

Fig. 2.16 (a) Two-stage fully balanced amplifier with common-mode feedback circuit, (b) the followed symbol of the amplifier

Table 2.5 Aspect ratios of the MOS transistors and component values of the fully balanced amplifier

Transistor	W/L (μm/μm)
Mn1, Mn5	50/1
Mn2, Mn4	200/1
Mn3	100/1
Mn6, Mn7	200/0.35
Mn8, Mn9	50/0.35
Mp1, Mp4	400/0.5
Mp2, Mp3	100/0.5
Mp5, Mp6	50/0.5
Component	Value
C_p	6 pF
R_p	3 kΩ
R_1, R_2	10 kΩ

that of MISO topology is followed in the case of SIMO universal biquad, where by employing an appropriate switching function, one of the five output currents is provided each time. The corresponding transfer function is derived according to the switch status condition of Table 2.6, while the implementation of the switching scheme is depicted in Fig. 2.17.

Table 2.6 Switch status condition for SIMO biquad

Control bits $b_2b_1b_0$	g_1	g_2	g_3	g_4	g_5	Filter function
000	1	0	0	0	0	LP
001	0	1	0	0	0	HP
010	0	0	1	0	0	BS
011	0	0	0	1	0	BP
100	0	0	0	0	1	AP

1: High-level; 0: Low-level

Fig. 2.17 Circuit scheme of filter function selection in SIMO universal biquad (DLC2)

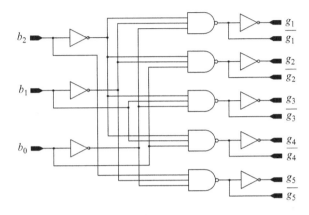

Table 2.7 Switch status condition for SIMO and MISO biquads

Control bits $b_3b_2b_1b_0$		Filter function
0000	MISO	LP
0001		HP
0010		BS
0011		BP
0100		AP
1000	SIMO	LP
1001		HP
1010		BS
1011		BP
1100		AP

Finally, it is obvious that an additional digital control bit is needed in order to select one of the ten transfer functions that are provided from the two universal biquads. Thus, the final filter function selection is performed by using the digital control bits $b_3b_2b_1b_0$ according to the switch status condition provided in Table 2.7.

The corresponding differential output current i_o, either from SIMO or MISO biquad, is provided by selecting the appropriate digital control bits and is converted into a voltage through the transresistance amplifier. The differential expression of the output voltage is given by $v_{0+} - v_{i-} = (i_{o+} - i_{o-}) \cdot R$. This procedure is illustrated in Fig. 2.18 where the output branches of SIMO and MISO universal biquads and the corresponding switches are shown.

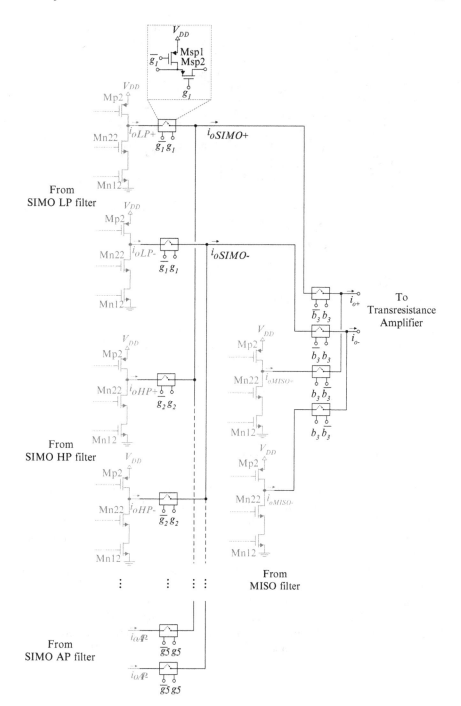

Fig. 2.18 Selection of the proper transfer function from SIMO and MISO universal biquads

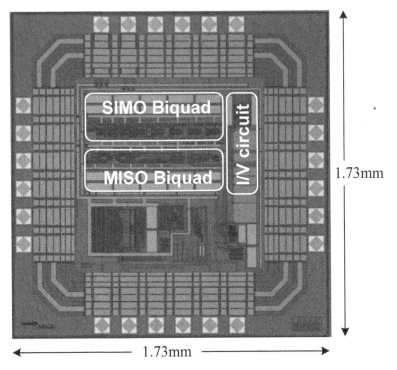

Fig. 2.19 Die microphotograph of the fabricated chip

2.4.5 Test Setup and Measurements

The die microphotograph of the fabricated chip is shown in Fig. 2.19. The employed general measurement setup is demonstrated in Fig. 2.20. The reference currents I_{01}, I_{02} are used to bias SIMO and MISO universal biquad, respectively. Each current is constructed by means of a low-voltage cascode current mirror and the external resistor R_{b1} and R_{b2}, respectively.

The Agilent 16315A balun transformer was used to convert a single-ended signal to the corresponding differential one, while the Agilent 1141A voltage differential probe was used to sense the outputs of the universal SIMO and MISO biquad filters. Extended measurements were taken using the HP4395 Spectrum Analyzer. The fabricated chip was tested in a PCB that was implemented through ORCAD PCB Editor 16.0 of Cadence. Photograph of the actual PCB is depicted in Fig. 2.21.

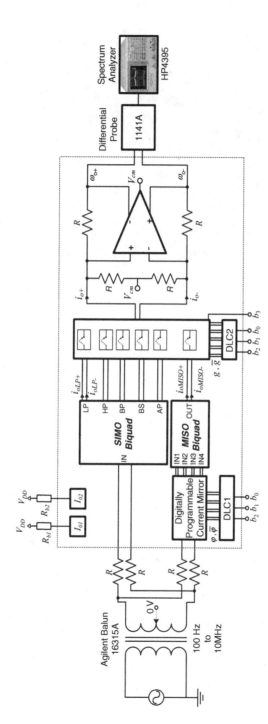

Fig. 2.20 General measurement setup for the current-mode SIMO and MISO universal biquads

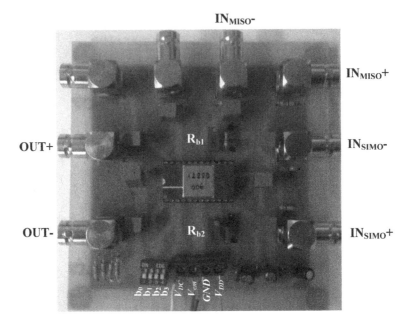

Fig. 2.21 Photograph of the PCB used to test the fabricated chip

The simulated and measured frequency responses of SIMO and MISO universal biquads are demonstrated in Fig. 2.22a, b, respectively. The measured Q values were 1.95 and 1.92 for the SIMO and MISO filter, respectively. These values are close to the nominal value of Q which is equal to 2. The capability of adjusting the resonant frequency ω_0 without disturbing the quality factor Q in both filters is depicted in Fig. 2.23, where I_0 was swept from 1.6 to 6 μA.

The measured center frequencies for $I_o = 1.6$, 2, 4 and 6 μA were 0.85, 1, 1.45 and 1.76 MHz, which are in good agreement with the theoretically predicted values 0.89, 1, 1.43 and 1.73 MHz.

Concerning the linearity of lowpass filters in SIMO and MISO biquads, two closely spaced tones (10 and 10.1 kHz) have been simultaneously applied at their inputs in order to have a −40 dB of third-order intermodulation distortion (IM3). The required summing operation has been performed by utilizing appropriately configured op-amps. The measured output spectrums of SIMO and MISO biquads are demonstrated in Fig. 2.24a, b, respectively. Considering the third-order inter-modulation products, the input power (P_{in}) was −11.11 dBm in the case of SIMO biquad resulting to an IIP3 of 8.88 dBm. In the case of MISO biquad, the input power was −10.26 dBm resulting to an IIP3 of 9.74 dBm, where IIP3 is calculated by (2.13).

The measured output noise densities of lowpass filters in SIMO and MISO biquads are shown in Fig. 2.25a, b, respectively. The input referred noise (*INOISE*) integrated over a 1 MHz range was −55.19 dBm for SIMO biquad and −52.74 dBm for MISO biquad. Using the (2.14), the values of SFDR for SIMO and MISO

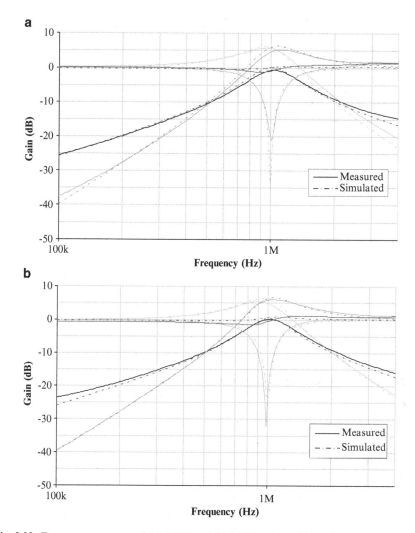

Fig. 2.22 Frequency responses of (**a**) SIMO and (**b**) MISO universal biquad

biquads were 42.7 and 41.6 dB, respectively. The observed values of dynamic range
are in good agreement with the corresponding simulated values of 44 and 42.6 dB,
respectively.

Finally, none of the 12 fabricated chips malfunctioned and thus the frequency
responses of all packaged samples could be measured. The measurement results for
both universal biquads and more specifically for the case of BP filter response are
shown in Fig. 2.26a, b, respectively. The center frequency as well as the Q factor
presents a very small variation between chips. However, any effect from process,
voltage and temperature variation can be eliminated by modifying the bias current
I_0, demonstrating thereby the attractive feature of electronic adjustment of filter
frequency characteristics.

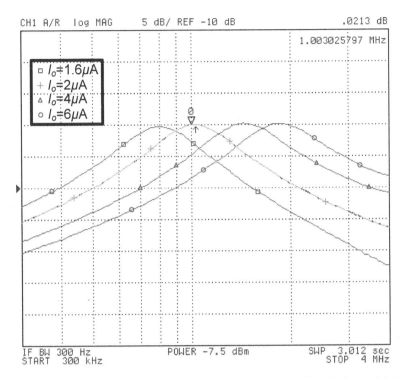

Fig. 2.23 Electronic adjustment of the resonant frequency in bandpass filter for $I_0 = 1.6, 2, 4$ and $6\,\mu$A

2.5 Summary

In this chapter novel SIMO and MISO universal biquads using low-voltage current mirrors have been presented. The proposed universal biquads provide the five standard filter functions while offering the feature of orthogonal adjustment between the resonant frequency and the quality factor. The behavior of the proposed filters is evaluated through simulation results, where the most important performance factors have been considered. The two topologies with the most optimum performance have been fabricated in AMS 0.35 μm C35B4C3 CMOS process and verified through extensive experimental measurements.

The proposed SIMO universal biquad topology has reduced power dissipation compared with the corresponding already published structures. Comparison results show that this is achieved without worsening their performance in terms of linearity, noise, dynamic range, and sensitivity. Experimental results also show that an IIP3 of 8.88 dBm is achieved, while the measured SFDR was 42.7 dB. With regards to the proposed MISO biquad topology, its performance is similar with that of the already introduced structure. Comparison results show that an almost 50%

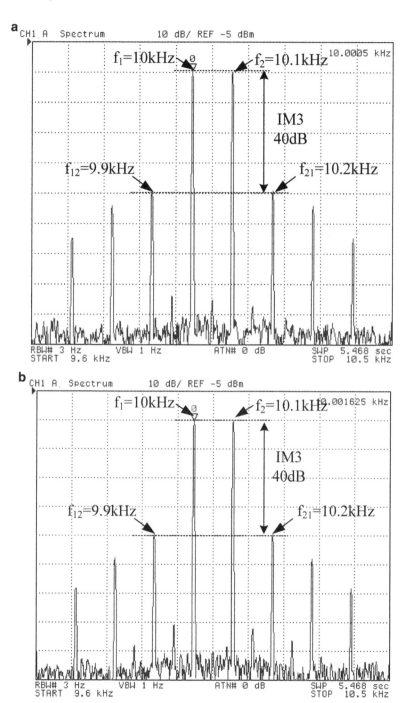

Fig. 2.24 Measured spectral plot for a two-tone test of the (**a**) SIMO and (**b**) MISO universal biquad

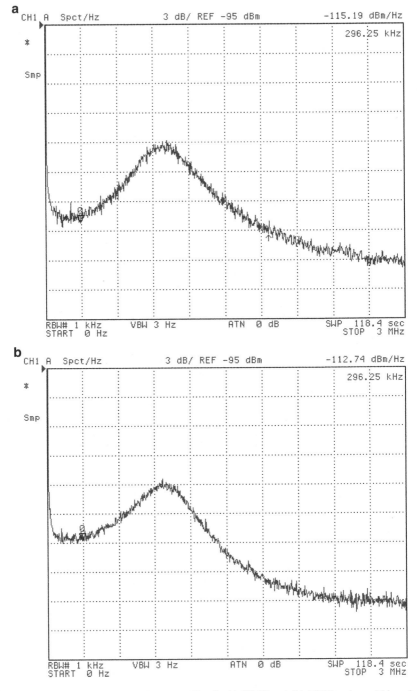

Fig. 2.25 Output noise density of lowpass filter in (**a**) SIMO and (**b**) MISO universal biquad

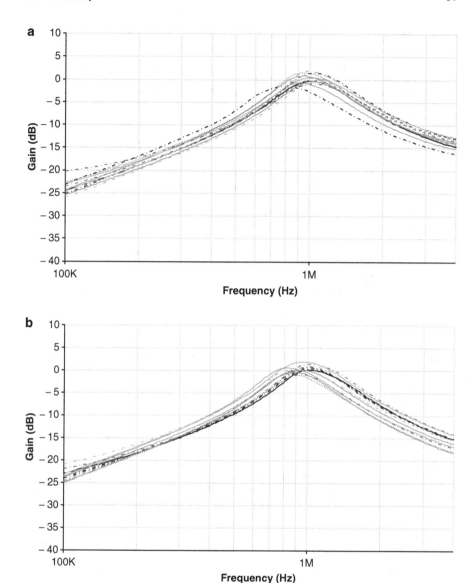

Fig. 2.26 Measured BP filter responses from the 12 packaged samples (**a**) SIMO and (**b**) MISO universal biquad

reduction of the required power dissipation is achieved. The measured IIP3 was 9.74 dBm and the SFDR was 41.6 dB.

From the aforementioned results, it is expected that the proposed topologies could be useful building blocks for realizing high-performance analog processing systems.

References

1. C. Toumazou, F. Lidgey, and D.G. Haigh, "Analog IC design: The current-node approach", Peter Peregrenus Ltd., London, UK, 1990.
2. J. Ramirez-Angulo, M. Robinson, and E. Sanchez-Sinencio, "Current-mode continuous-time filters: two design approaches", *IEEE Transactions on Circuits and Systems II*, vol. 39, no. 6, pp. 337–341, Jun. 1992.
3. S.S. Lee, R.H. Zele, D.J. Allstot, and G. Liang, "CMOS Continuous-Time current-mode filters for high frequency applications", *IEEE Journal of Solid-State Circuits*, vol. 28, no. 3, pp. 323–329, Mar. 1993.
4. A.H.M. Shousha, "Implementations of continuous-time current-mode ladder filters using multiple output current integrators", *International Journal of Electronics*, vol. 85, no. 4, pp. 497–509, 1998.
5. G. Souliotis, A. Chrisanthopoulos, and I. Haritantis, "Current Differential Mirrors: New circuits and applications", *International Journal of Circuit Theory Applications*, vol. 29, no. 6, pp. 553–574, Nov./Dec. 2001.
6. G. Souliotis, and C. Psychalinos, "Harmonic oscillators realized using current amplifiers and grounded capacitors", *International Journal of Circuit Theory and Applications*, vol. 35, no. 2, pp. 165–173, Mar. 2007.
7. G. Souliotis, and I. Haritantis, "Current-mode filters based on current mirror arrays", *International Journal of Circuit Theory and Applications*, vol. 36, no. 2, pp. 173–183, Mar. 2008.
8. G. Souliotis, and C. Psychalinos, "Electronically controlled multiphase sinusoidal oscillators using current amplifiers", *International Journal of Circuit Theory and Applications*, vol. 37, no. 1, pp. 43–52, Feb. 2009.
9. M.A. Ibrahim, S. Minaei, and H. Kuntman, "A 22.5 MHz current-mode KHN-biquad using differential voltage current conveyor and grounded passive elements", *International Journal of Electronics and Communications (AEU)*, vol. 59, no. 5, pp. 311–318, Jul. 2005.
10. L.D. Paarmann, "Design and analysis of analog filters. A signal processing respective", Kluwer Academic Publishers: Dordrecht, ch.1, 2001.
11. A.M. Soliman, "Voltage mode and current mode Tow Thomas biquadratic filters using inverting CCII", *International Journal of Circuit Theory and Applications*, vol. 35, no. 4, pp. 463–467, Jul. 2008.
12. S.I. Liu, J.J. Chen, and Y.S. Hwang, "New current mode biquad filter using current followers", *IEEE Transactions on Circuits and Systems I*, vol. 42, no. 7, pp. 380–383, Jul. 1995.
13. M. Okine, and N. Katsuhara, "Current-mode active RC filters using current followers", *Analog Integrated Circuits and Signal Processing*, vol. 20, no. 2, pp. 161–164, Aug. 1999.
14. S. Celma, J. Sadabell, and P. Martinez, "Universal filter using unity-gain cells", *Electronics Letters*, vol. 31, no. 21, pp. 1817–1818, Oct. 1995.
15. M.T. Abuelma'atti, M.A. Al-Qahtani, "Current-mode universal filters using unity-gain cells", *Electronics Letters*, vol. 32, no. 12, pp. 1077–1079, Jun. 1996.
16. E.O. Gunes, F. Anday, "Realization of voltage and current-mode transfer functions using unity-gain cells", *International Journal of Electronics*, vol. 83, no. 2, pp. 209–213, Aug. 1997.
17. H.A. Alzaher, M. Ismail, "Current-mode universal filter using unity-gain cells", *Electronics Letters*, vol. 35, no. 25, pp. 2198–2200, Dec. 1999.
18. R.M. Weng, J.R. Lai, M.H. Lee, "New universal biquad filters using only two unity-gain cells", *International Journal of Electronics*, vol. 87, no. 1, pp. 57–61, Jan. 2000.
19. C.M. Chang, T.S. Liao, T.Y. Yu, E.S. Lin, C.H. Teng, and C.L. Hou, "Novel universal current-mode filters using unity-gain cells", *International Journal of Electronics*, vol. 82, no. 1, pp. 23–30, 2001.
20. V.K. Singh, A.K. Singh, D.R. Bhaskar, and R. Senami, "New universal biquads employing CFOAs", *IEEE Transactions on Circuits and Systems II*, vol. 53, no. 11, pp. 1299–1303, Nov. 2006.

21. H.P. Chen, Y.Z. Liao, and W.T. Lee, "Tunable mixed-mode OTA-C universal filter", *Analog Integrated Circuits and Signal Processing*, vol. 58, no. 2, pp. 135–141, Feb. 2009.
22. M.T. Abuelma'atti, and A. Bentrcia, "A novel mixed-mode OTA-C universal filter", *International Journal of Electronics*, vol. 92, no. 7, pp. 375–383, 2005.
23. W. Tangsrirat, "Low-voltage digitally programmable current-mode universal biquadratic filter", *International Journal of Electronics and Communications (AEU)*, vol. 62, no. 2, pp. 97–103, Feb. 2008.
24. C. Laoudias, C. Psychalinos, "Single Input Multiple Output Universal Biquad Using Current Mirrors", in *Proc. of IEEE International Conference on Electronics, Circuits and Systems* (ICECS), Malta, pp.1026–1029, Aug. 2008.
25. C. Laoudias, C. Psychalinos, "Universal Biquad Filters Using Low-Voltage Current Mirrors", *Analog Integrated Circuits and Signal Processing*, vol. 65, no. 1, pp. 77–88, Oct. 2010.
26. C. Laoudias, C. Psychalinos, "A Low-Voltage Current-mode Single Input Multiple Output Universal Biquad Filter", in *Proc. of IFIP/IEEE International Conference on Very Large Scale Integration* (VLSI-SoC), Rhodes, pp. 497–500, Oct. 2008.
27. C. Laoudias, C. Psychalinos, "Universal Biquad Filter Topology Using Low-Voltage Current Mirrors", *International Journal of Circuits Theory and Applications*, DOI: 10.1002/cta.706.
28. C. Laoudias, C. Psychalinos, "Multiple Input Multiple Output Current-Mode Universal Biquad Filters", in *Proc. of 15th IEEE Mediterranean Electrotechnical Conference* (MELECON), Malta, pp. 296–299, Apr. 2010.
29. E. Sackinger, and W. Guggenbuhl, "A high-swing high-impedance MOS cascode circuit", *IEEE Journal of Solid-State Circuits*, vol. 25, no. 1, pp. 289–298, Feb. 1990.
30. H. Zouaoui-Abouda, A. Fabre, "A New Balanced CMOS Controlled Integrator for Ultra High Frequency Applications", *Analog Integrated Circuits and Signal Processing*, vol. 47, no. 1, pp. 13–22, Apr. 2006.
31. S.L. Smith, and E. Sanchez-Sinencio, "Low Voltage Integrators for High-Frequency CMOS Filters Using Current Mode Techniques", *IEEE Transactions on Circuits and Systems I*, vol. 43, no. 1, pp. 41–48, Jan. 1996.

Chapter 3
Complex Filters for Short Range Wireless Networks

Abstract The design of low-voltage complex filters using current mirrors for wireless receivers is presented in this chapter. Complex signal processing is an attractive technique for removing the image signals that appear in transceiver architectures. The problem from the presence of image signal in low-IF architectures is caused by the down-conversion operation realized by complex mixing. The realization of complex filters is achieved by employing an appropriate transformation to the corresponding conventional real filters. Two systematic methods for designing high-order complex filters are presented i.e. the leapfrog and the topological emulation techniques, where the employed active elements are low-voltage current mirrors. Thus, like in previous chapter, the offered benefits are the capability of low-voltage operation, the electronic tuning of their frequency responses and the absence of resistors. A twelfth-order complex filter function has been realized by employing the aforementioned techniques, where the performance of the corresponding topologies, fabricated in AMS 0.35 μm CMOS process, has been experimentally verified. Also, a detailed test setup for the measurement of the fabricated chip, including the interface for V/I and I/V conversion, is also provided.

Keywords Bluetooth/ZigBee low-IF receivers • Complex signal processing • Current mirror filters • Current mode circuits • Image rejection • Short range wireless networks

3.1 Introduction

The fast growth of wireless applications in recent years has driven intense efforts to design low-power, low-cost and highly integrated transceivers. The increasing demand of portable devices, especially with the capability of operating in various standards, has fuelled the use of short range wireless personal networking (Bluetooth, 802.11b/a/g, ZigBee). The design of wireless transceivers is particularly a challenging procedure in order to achieve the highest level of integration,

C. Laoudias and C. Psychalinos, *Integrated Filters for Short Range Wireless and Biomedical Applications*, SpringerBriefs in Electrical and Computer Engineering, DOI 10.1007/978-1-4614-0260-2_3, © Springer Science+Business Media, LLC 2012

lowest power consumption, while aiming to achieve the most optimum performance. Of course, all these requirements are not met in a single architecture, and therefore, several tradeoffs must be made to find the best architecture that meets the standard specifications. Moreover, the choice of the most suitable receiver architecture depends on various parameters of the wireless standard such as channel bandwidth, sensitivity, selectivity, blocking specifications.

The most preferred approach in fully integrated wireless receivers operating in Bluetooth standard is the low-IF (Intermediate Frequency) architecture. Low-IF architectures can be used to avoid the DC offset and 1/f noise (flicker noise) associated with direct-conversion. However, these structures suffer from the image problem caused by the down-conversion operation. Unfortunately, the conventional real filters have not the capability for removing these undesired signals due to their symmetrical response around dc. In order to overcome this problem, a new class of filters denoted as complex filters has been introduced in the literature. Complex filters are constructed from two-path networks, where a pair of signals with equal amplitudes and quadrature phases (I and Q channels) is applied at their inputs. The concept of complex signal processing is formally described in [1–3].

A significant research effort has been already performed in the literature for designing complex filters suitable for low-IF receivers. The topologies in [4, 5] have been derived using the switched-capacitor and switched-current techniques, respectively, while the topologies in [6, 7] are using the concept of companding signal processing and bipolar transistors. The topologies in [8–15] are derived by employing the concept of conventional linear continuous-time filtering and MOS transistors. The active elements in [8–12] are the Operational Transconductance Amplifiers (OTAs). Second generation Current Conveyors (CCIIs) have been utilized in [13]. CCIIs configured as Current Followers (CFs) and Voltage Followers (VFs) have been employed in [14, 15]. Among the already presented topologies, only these in [8–12] offer the capability of resistorless realization, while these in [13–15] could be resistorless by substituting passive resistors by MOS transistors in triode region. In addition, the circuit complexity in [8–12] is smaller than that of the circuits in [13–15] due to the fact that OTAs have simpler structure than CCIIs, CFs and VFs. Also, the employed power supply voltages were 2.7 V in [10], [14, 15], 2.5 V in [8, 9] and 1.8 V in [11].

Another alternative element for realizing resistorless filters with electronic tuning capability is the current mirror. The internal structure of a current mirror is simpler than that of OTA, making the current mirror an attractive element for performing analog signal processing. Current mirrors have been widely used for realizing real filter functions [16, 17].

Novel complex filter topologies using low-voltage current mirrors are introduced in this chapter, where it is organized as follows: in the first part the complex filter theory is presented. In the second part, a twelfth-order complex filter is designed using the Leapfrog method in order to meet the Bluetooth/ZigBee standard requirements. Finally, in the third part the Leapfrog and the topological emulation of the corresponding passive filters are utilized in order to design two twelfth-order complex filters that fulfill the specifications of Bluetooth standard.

The two complex filters have been fabricated using AMS 0.35 μm C35B4C3 CMOS process and the obtained experimental results confirm their correct operation.

3.1.1 Design Considerations of Bluetooth Receiver Architectures

There are two main types of receivers used in recent telecommunication systems based on phase or frequency modulation: the heterodyne and the homodyne receiver. The difference between them is in whether or not an IF is used for the down-conversion of the Radio Frequency (RF) signal. In heterodyne receivers, the RF signal is down-converted from its carrier to a low or high IF, resulting into two different subcategories, the low-IF and high-IF, respectively. In homodyne receivers, the signal is down-converted directly from RF to baseband and for that reason they are also called as direct conversion or zero-IF receivers.

The choice of the IF affects the design of the whole system and results into different design characteristics including sensitivity, selectivity, channel bandwidth, blocking specifications and channel selection filters. Therefore, one of the most challenging parts of the receiver is the channel filtering, which is discussed thoroughly in this chapter. In the following, a comparison between these types of receivers is performed in order to find the most suitable architecture for the Bluetooth/ZigBee wireless standards.

A high-IF receiver, which uses an IF much larger than the signal channel bandwidth, requires off-chip components with high quality factors (Q); hence, the system integration level is reduced, and extra power on the I/O driving circuits is demanded. In addition, the high-IF choice also increases the complexity of the IF band circuits and causes more power dissipation in the IF stage. On the other hand, low-IF receivers circumvent the previously mentioned problems, where a conversion to an IF that is near the channel bandwidth is used. Thus, high order filters but with lower quality factors are required for the channel selection. This scheme provides a fully integrable and low-power solution. However, the main disadvantage of heterodyne receivers is the presence of image signal, caused by the down-conversion of the desired RF signal. For this reason, a specific form of filters called complex filters is required for the image rejection. In contrary to the heterodyne receivers, the direct conversion receivers do not suffer from the presence of image signal due to the fact that the IF is equal to zero. So, the channel selection filtering is performed easier, where only a lowpass filter with relatively sharp cutoff characteristics is needed. Though, the direct conversion of RF signal entails a number of issues that are not critical or even do not exist in a heterodyne receiver. The main drawback of direct conversion receivers is the presence of flicker noise which causes significant degradation in the Signal-to-Noise Ratio (SNR). Also, direct conversion receivers suffer from DC offsets mainly caused by self-mixing. This problem is originated from the fact that the isolation between the ports of Local Oscillator (LO) and the inputs of mixer and LNA is not infinite. That is, a portion of the signal from the LO appears at the inputs of the LNA, where it is amplified and then mixed with the LO signal, thus producing a DC component. The problem of DC offsets can be alleviated using various techniques, leading though to higher complexity and increased power consumption.

From the above discussion and considering that in a Bluetooth signal, the 99% of the signal power is contained within the DC to 430 kHz, then due to flicker noise and DC offsets the selection of direct conversion receiver would lead to a degraded SNR. Consequently, the low-IF architecture is the most suitable architecture for Bluetooth, especially when considering the relaxed image rejection requirement of the Bluetooth standard. In the next section, the problem of the presence of image signal in low-IF receivers and the image rejection from the complex filters are analyzed.

3.1.2   The Problem of Image signal in Low-IF Receivers

Low-IF receivers have been in use for a long time and their principle of operation is well known. In order to understand the problem from the presence of image signal, it must at first be explained the down-conversion of the desired signal into IF. This operation is realized by the mixer where the desired signal $x_{RF}(t)$ is multiplied with the signal from the LO $x_{LO}(t)$ and thus the signal at the output of mixer ω_{IF} will be located at a lower frequency (ω_{IF}), as shown in Fig. 3.1.

The main problem is that except from the desired signal, the unwanted image signal, located at a frequency ω_{IM}, which is $2\omega_{IF}$ away from the wanted signal, is down-converted to the same IF. Considering the following signals, then a mathematical description of the image frequency problem can be expressed as:

$$x_{RF}(t) = 2\cos\omega_{RF} = e^{j\omega_{RF}t} + e^{-j\omega_{RF}t} \tag{3.1a}$$

$$x_{IM}(t) = 2\cos\omega_{IM} = e^{j\omega_{IM}t} + e^{-j\omega_{IM}t} \tag{3.1b}$$

$$x_{LO}(t) = 2\cos\omega_{LO} = e^{j\omega_{LO}t} + e^{-j\omega_{LO}t} \tag{3.1c}$$

Also, without loss of generality, by assuming that the signals have the following relationships:

$$\left.\begin{array}{c} \omega_{RF} - \omega_{LO} = \omega_{IF} \\ \omega_{LO} - \omega_{IM} = \omega_{IF} \end{array}\right\} \tag{3.2}$$

then the mixer output y(t) is given by[1]

---

[1] The amplitude of each signal has been ignored, since the frequency components are the subject to be discussed.

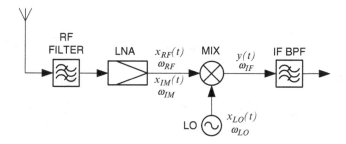

Fig. 3.1 Block diagram of the front-end stage in a low-IF receiver

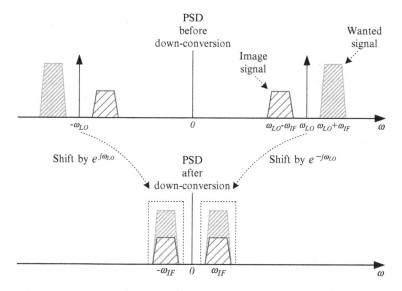

Fig. 3.2 Spectral densities of the signals before and after the down-conversion with a real LO signal

$$y(t) = [x_{RF}(t) + x_{IM}(t)] \cdot x_{LO}(t) \Rightarrow$$
$$y(t) = e^{j(\omega_{RF}+\omega_{LO})t} + e^{j(\omega_{RF}-\omega_{LO})t}$$
$$+ e^{-j(\omega_{RF}-\omega_{LO})t} + e^{-j(\omega_{RF}+\omega_{LO})t}$$
$$+ e^{j(\omega_{IM}+\omega_{LO})t} + e^{j(\omega_{IM}-\omega_{LO})t}$$
$$+ e^{-j(\omega_{IM}-\omega_{LO})t} + e^{-j(\omega_{IM}+\omega_{LO})t} \tag{3.3}$$

The spectral densities of the signals before and after the down-conversion with the real signal from LO are depicted in Fig. 3.2, where. Equation 3.3 reveals that the input RF spectrum is down-converted to an IF and up-converted to a higher RF. With dashed box is represented the real bandpass filter response. In a receiver, the up-converted components are not a problem if the mixing operation is followed by

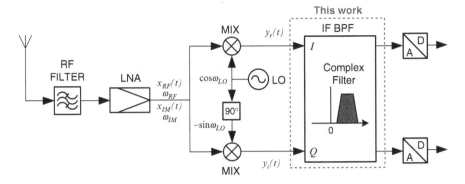

Fig. 3.3 Block diagram of the front-end stage in a low-IF Bluetooth receiver using a complex filter

lowpass filtering. More important is the fact that the RF frequency components located at $\omega_{LO} + \omega_{IF}$ and $\omega_{LO}-\omega_{IF}$ are both converted to the same IF. To avoid corrupting the desired signal, the image signal must be suppressed before down conversion by means of a bandpass RF filter. The quality factor of such filter is proportional to f_{RF}/f_{IF}. High quality factor external SAW or ceramic filters are used for this purpose, increasing thus the complexity and the power consumption of the whole receiver.

Therefore, the overlap of the two spectra is caused by the mixing operation with a real signal which has two frequency components of $\pm\,\omega_{LO}$.

An obvious solution to avoid this problem is to multiply the RF signal with a LO signal which has only one frequency component, say $e^{-j\omega_{LO}}$. Such image rejection system is achieved by using complex signal processing. To implement this complex multiplication using real components, two signal branches must be constructed, namely I and Q. In the I (in-phase or real) branch, the RF signal is multiplied by $\cos\omega_{LO}$, while in the Q (quadrature phase or imaginary), the RF signal is multiplied by $-\sin\omega_{LO}$. Figure 3.3 shows the block diagram of a low-IF Bluetooth receiver using a complex filter.

Let the LO signal be a complex sinusoid expressed by

$$\mathbf{v}_{LO}(t) = e^{-j\omega_{LO}t} = \cos\omega_{LO}t - j\sin\omega_{LO} \tag{3.4}$$

Considering that the signals (3.1a) and (3.1b) are inserted in the mixer and multiplied by the complex signal (3.4), then the mixer output $y(t)$ is now given by

$$\mathbf{y}(t) = [x_{RF}(t) + x_{IM}(t)] \cdot \mathbf{v}_{LO}(t) \Rightarrow$$
$$\mathbf{y}(t) = e^{j(\omega_{RF}-\omega_{LO})t} + e^{-j(\omega_{RF}+\omega_{LO})t} + e^{j(\omega_{IM}-\omega_{LO})t} + e^{-j(\omega_{IM}+\omega_{LO})t} \tag{3.5}$$

where $y(t)$ is the complex signal $\mathbf{y}(t) = y_r(t) + jy_i(t)$. By ignoring the terms with the sum of frequencies, for the same reason as in previous case, and considering that

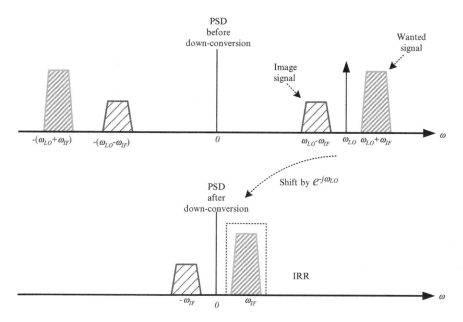

Fig. 3.4 Spectral densities of the signals before and after the down-conversion with a complex LO signal

x_{sig} and x_{im} are the amplitudes of the signals $x_{RF}(t)$ and $x_{IM}(t)$, respectively, then (3.5) can be rewritten as

$$
\begin{aligned}
\mathbf{y}(t) &= \mathrm{x}_{\mathrm{sig}} e^{j(\omega_{RF}-\omega_{LO})t} + \mathrm{x}_{\mathrm{im}} e^{j(\omega_{IM}-\omega_{LO})t} \overset{(3.2)}{=} \\
&= \mathrm{x}_{\mathrm{sig}} e^{j\omega_{IF}t} + \mathrm{x}_{\mathrm{im}} e^{-j\omega_{IF}t} = \\
&= \mathrm{x}_{\mathrm{sig}}(\cos\omega_{\mathrm{IF}} + j\sin\omega_{IF}) + \mathrm{x}_{\mathrm{im}}(\cos\omega_{\mathrm{IF}} - j\sin\omega_{IF})
\end{aligned}
\tag{3.6}
$$

From (3.6), it is revealed that the desired and the complex signals are explicitly separated, where the former is located at ω_{IF}, while the latter is at $-\omega_{IF}$. The spectral densities of the signals before and after the down-conversion with a complex LO signal is given in Fig. 3.4, where the complex bandpass filter response is denoted with a dashed box. Thus, the image signal can be rejected by means of a complex filter, while the same topology is used for channel selectivity. A benefit of complex filters is that the quality factor of such filters is proportional to ω_{IF}/BW, where BW is the channel bandwidth and it has small values in low-IF receivers. However, complex filters are very sensitive to I/Q mismatches, i.e. the phase and gain imbalances at the mixer output, due to LO and mixer mismatches, resulting to a limited image rejection ratio (IRR). In the case of a phase shift of 90° and/or amplitude mismatch between the signals in I and Q branches, some power will leak from the image band into the signal band and vice-versa. Therefore, as it is shown in Fig. 3.5, the IRR quantifies the relation between the negative half-plane image band and positive half-plane leaked image power in low-IF receivers

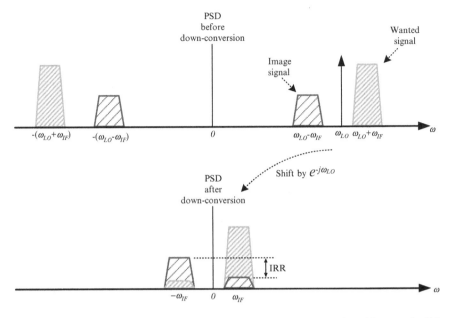

Fig. 3.5 Spectral densities of the signals before and after the down-conversion with a complex LO signal, including the effect of I/Q mismatches.

After a lot of algebraic operations, it can be proved [14] that IRR is given by

$$IRR = \frac{1 - 2(1 + \varepsilon)\cos\theta + (1 + \varepsilon)^2}{1 + 2(1 + \varepsilon)\cos\theta + (1 + \varepsilon)^2} \tag{3.7}$$

where ε denotes the relative gain mismatch and θ the phase imbalance between I and Q branches. For small values of ε and θ, (3.7) reduces to

$$IRR = \frac{\varepsilon^2 + \vartheta^2}{4} \tag{3.8}$$

where θ is in radians. For example, if $\varepsilon = 5\%$ and $\theta = 5°$, then IRR ≈ 26 dB. In the case of IRR $= 60$ dB, then θ must remain below $0.1°$, a value difficult to attain in typical IC technologies. In practice, these type of filters exhibit an IRR of 20–30 dB, as it will be shown in the next sections where are designed complex filters using current mirrors.

3.1.3 Complex Filter Theory

The conventional real filters have not the capability to remove the image signal due to their symmetrical response around dc. On the other hand, complex filters have an

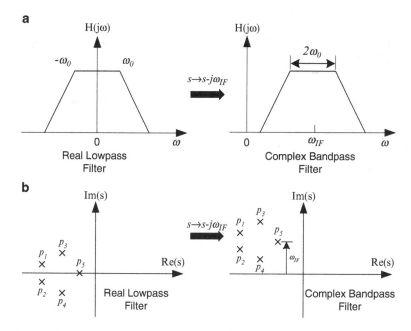

Fig. 3.6 Frequency shifting of a real lowpass filter. Effect on (**a**) transfer function and (**b**) pole locus

unsymmetrical frequency response around the jω axis, though they are symmetrical around ω_{IF}. That is, complex filters can pass the desired signal at $\omega = \omega_{IF}$, while attenuating the image at $\omega = -\omega_{IF}$. So, the principle of the complex filters is that, in the complex domain the complex bandpass filter is a frequency shifted version of a real lowpass filter. This frequency shifting is realized by employing the following transform

$$s \rightarrow s - j\omega_{IF} \qquad (3.9)$$

where ω_{IF} denotes the shift of the transfer function. Applying this translation in the transfer function of a real lowpass filter $H_{LP}(s)$, then the transfer function of the complex bandpass filter $H_{BP}(s)$ is given by (3.10).

$$H_{BP}(s) = H_{LP}(s - j\omega_{IF}) \qquad (3.10)$$

Moreover, this translation remaps all poles in the s-plane by moving them from their position located around the zero axis to the center frequency ω_{IF}. This is exemplified in Fig. 3.6, where the case of a fifth-order filter has been considered. In practice, the transformation of a real lowpass filter into a complex bandpass filter is realized by shifting in frequency any frequency dependent element of it. In other words, (3.9) must be applied to any lossy and lossless integrator of the active filter.

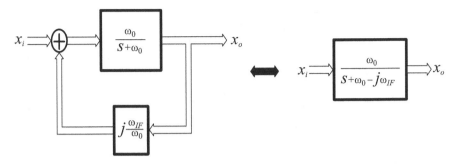

Fig. 3.7 Diagram for the frequency shifting of a real lossy integrator

Consider the simple case of converting a real lossy integrator, where its transfer function is

$$H_{LP}(s) = \frac{\omega_0}{s + \omega_0} \tag{3.11}$$

to a complex lossy integrator centered at ω_{IF}, where its transfer function is given by

$$H_{BP}(s) = \frac{\omega_0}{s - j\omega_{IF} + \omega_0} \tag{3.12}$$

Then, the lowpass transfer function of (3.11) is shifted in frequency by placing it in a complex feedback loop as shown in Fig. 3.7. Considering that $x_i = x_{iI} + jx_{iQ}$ and $x_o = x_{oI} + jx_{oQ}$ are the input and output complex signals, respectively, then (3.12) can be rewritten as

$$x_o = \frac{\omega_0}{s - j\omega_{IF} + \omega_0} x_i \tag{3.13}$$

From (3.13) it is revealed that

$$x_{oI} = \frac{\omega_0}{s + \omega_0} \left(x_{iI} - \frac{\omega_{IF}}{\omega_0} x_{oQ} \right) \tag{3.14a}$$

$$x_{oQ} = \frac{\omega_0}{s + \omega_0} \left(x_{iQ} + \frac{\omega_{IF}}{\omega_0} x_{oI} \right) \tag{3.14b}$$

Equation 3.14 is implemented as shown in Fig. 3.8.

From the FBD in Fig. 3.8, it is evident that operations like addition, scaling and inversion are very easily realized by utilizing current mirrors. Thus, the concept introduced in the following sections is based on the employment of low-voltage

Fig. 3.8 FBD of a complex lossy integrator

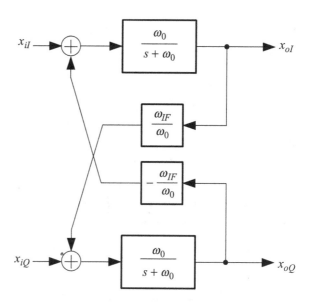

current mirrors in order to realize complex filters for Bluetooth receivers. For this purpose, the leapfrog and topological emulation of the corresponding passive prototypes have been performed.

3.2 Complex Filter Design Using Leapfrog Method

In this section, the design of complex filters using the Leapfrog method is described. Leapfrog method is a very common method for the synthesis of high-order filters and is based in the operational emulation of LC ladder prototypes. According to this approach, the voltage-current relationships of LC ladder prototype are expressed in a way that all variables are currents. The required current operations are implemented using lossy and lossless integrators constructed by current mirrors. The result from this step is the derivation of the FBD of the filter using real integrators. One benefit of the realized topologies is that the addition of the currents is implemented very easily by connecting the corresponding outputs. The next step is to substitute the real integrators of the conventional filter's FBD by the corresponding complex integrators. According to the previous section, this is readily obtained by applying (3.9) to each real lossy and lossless integrator. The derivation of complex integrators using current mirrors will be presented in the next section. The above procedure is illustrated in Fig. 3.9, where ω_{0i} ($i = 1,2,\dots n$) is the characteristic frequency of each integrator, ω_{IF} is the frequency shifting and $i_{i,c} = i_{iI} + ji_{iQ}$, $i_{o,c} = i_{oI} + ji_{oQ}$ are the input-output complex signals in I and Q branches, respectively.

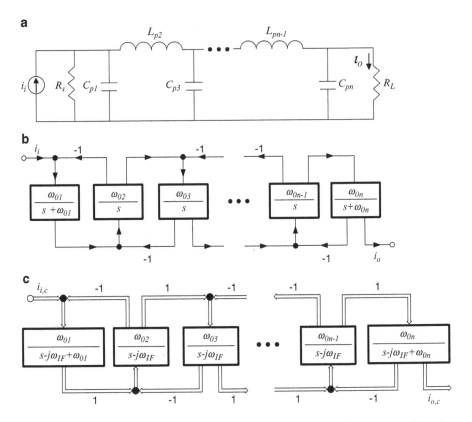

Fig. 3.9 Synthesis of high-order filters using Leapfrog method (**a**) Passive prototype of an nth-order lowpass filter (**b**) FBD of an nth-order real lowpass filter (**c**) FBD of nth-order complex bandpass filter

3.2.1   Complex Iintegrators Using Current Mirrors

The topologies of lossy and lossless integrators using current mirrors have been presented in the previous chapter. The realized transfer functions have the forms given by (3.15) and (3.16), respectively

$$H_1(s) = \frac{i_o}{i_i} = \frac{\omega_0}{s + \omega_0} \tag{3.15}$$

$$H_2(s) = \frac{i_o}{i_i} = \frac{\omega_0}{s} \tag{3.16}$$

where $\omega_0 = 1/\tau$. The time constant in both transfer functions is given by the expression $\tau = C/g_m$, where g_m is the small-signal transconductance parameter of the transistor at the input of the current mirror.

Fig. 3.10 Realization of complex (**a**) lossy and (**b**) lossless integrator using current mirrors

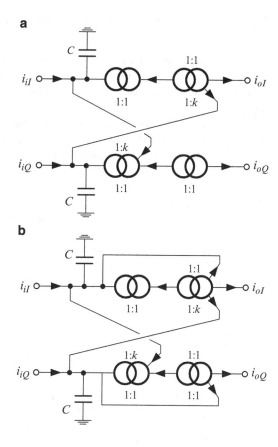

The above transfer functions are transposed to the corresponding ones in the complex domain by applying the frequency transformation of (3.9). Thus, the transfer functions of complex lossy and lossless integrators are given by (3.17) and (3.18), respectively.

$$H_1(s) = \frac{i_o}{i_i} = \frac{\omega_0}{s + \omega_0 - j\omega_{IF}} \tag{3.17}$$

$$H_2(s) = \frac{i_o}{i_i} = \frac{\omega_0}{s - j\omega_{IF}} \tag{3.18}$$

The realizations of complex integrators using current mirrors are depicted in Fig. 3.10, where the scaling factor k is the current gain of the corresponding output and is expressed by

$$k = \frac{\omega_{IF}}{\omega_0} = \omega_{IF}\frac{C}{g_m} \tag{3.19}$$

3.2.2 Complex Filter for Bluetooth/ZigBee Standards

As a design example, a complex filter that will meet the requirements of Bluetooth/ ZigBee standards will be presented [15]. The complex filter is used for the channel selection and image rejection and will operate at a different standard each time by applying an appropriate switching scheme. The most critical requirements of the two standards, concerning the performance of the complex filter, are IRR, blocker attenuation, linearity and the in-band group delay variation.

Bluetooth technology was developed to create a short-range wireless voice and data link between a broad range of devices such as PCs, notebook computers, PDAs, mobile phones and digital cameras. Consistent with its aim of operating in even the smallest battery-powered devices, the Bluetooth specification calls for a small form factor, low power consumption and low cost. The range and speed of the technology were kept intentionally low so as to ensure maximum battery life and minimum incremental cost for devices incorporating the technology. A Bluetooth transceiver is a frequency-hopping spread-spectrum device that uses the unlicensed (worldwide) 2.4 GHz ISM (Industrial, Scientific, Medical) frequency band. There are 79 channels available, where the nominal bandwidth for each channel is 1 MHz, while the effective data rate is at 723.2 Kbit/s.

ZigBee is another one low-cost, low-power, short range wireless standard. It operates either on ISM band using 16 channels with 5 MHz bandwidth per channel or at the lower frequency band of 902–828 MHz, using ten channels with 2 MHz bandwidth per channel.

Since these two standards are operating in the same frequency band, then the RF front-end of the low-IF receiver can be shared between the two operating modes. About the operation of the complex filter in Bluetooth standard, the filter bandwidth must be 1 MHz, i.e. equal to the channel bandwidth, while in ZigBee mode the filter bandwidth must be equal to 2 MHz. The center frequency ω_{IF} of the complex filter is chosen equal to 1 MHz for the Bluetooth mode, while in the ZigBee mode is chosen equal to 2 MHz. Other requirements of the two wireless standards with regards to the complex filter design are: (1) Blocking attenuation of adjacent channels, where in Bluetooth standard, the complex filter must exhibit 11, 41 and 51 dB attenuation at 1, 2 and 3 MHz away from the center frequency, while the same attenuation holds about ZigBee mode, for frequencies 2, 4 and 6 MHz away from the center frequency, (2) IRR, where 20–30 dB is adequate in both standards and (3) In-band group delay variation, where it must be ≤ 1 μs in both standards.

System level simulations show that a complex filter based on a sixth-order lowpass Butterworth filter with a cutoff frequency (ω_0) 500 kHz is sufficient to achieve the aforementioned specifications. Thus, in Bluetooth standard the complex filter will have 1 MHz center frequency and a bandwidth of 500 kHz on each side of the center frequency, while in ZigBee standard the center frequency and the bandwidth are exactly twice than those in Bluetooth mode. Consequently, the ZigBee requirements can be fulfilled either by doubling the dc bias current of the current mirrors or by utilizing an appropriate switching scheme and extra capacitors in order to half the value of integration capacitances. However, the former solution implies a significant increase of power consumption, whereas the latter has an increased total capacitor area. In the current design example, the second solution was chosen in order to keep the power consumption low.

Using the FBD in Fig. 3.9 and the integrator topologies in Fig. 3.10, the derived complex filter is that given in Fig. 3.11. In order to fulfill the nowadays trend for realizing analog signal processing systems capable of operating in low-voltage environment, the current mirror shown in Fig. 2.6 has been employed in the design of the filter.

The behavior of the complex filter in Fig. 3.11 is evaluated through simulation results, by employing the Virtuoso Analog Design Environment of the Cadence software. Level 49 MOS transistor models of the AMS 0.35 μm CMOS process have been used in simulations. Considering a dc bias current $I_0 = 16.4$ μA and supply voltages $V_{DD} = 1.5$ V and $V_{DC} = 1.2$ V, the values of the NMOS transistors aspect ratios were 5 μm/1 μm for M_{n1i} ($i = 1, 2, ...$) and 7 μm/0.6 μm for M_{n2i} ($i = 1, 2, ...$). The PMOS transistors have an aspect ratio 15 μm/2 μm. Under the above bias conditions, the input resistance of the current mirrors was $R_{in} = 1/g_{m}$, $M_{n11} = 7.45$ kΩ. Thus, the values of capacitors and current gains are calculated according to the formulas provided by

$$C_i = C_{pi}, (i = 1, 3, 5) \tag{3.20a}$$

$$C_j = \frac{L_{pj}}{R_{in}^2}, (j = 2, 4, 6) \tag{3.20b}$$

where C_{pi} and L_{pj} are the component values of the passive prototype. The capacitor values of the passive prototype as well as the complex filter for both standards are summarized in Table 3.1.

According to (3.19)–(3.20), the denoted current gains k_i and k_j in Fig. 3.11 are expressed by

$$k_i = \omega_{IF} R_{in} C_{pi}, (i = 1, 3, 5) \tag{3.21a}$$

$$k_j = \omega_{IF} \frac{L_{pj}}{R_{in}}, (j = 2, 4, 6) \tag{3.21b}$$

The simulated Bluetooth and ZigBee frequency responses for the signal and image sides are given in Fig. 3.12. Obviously, the signal response is achieved by using two quadrature signals at I and Q inputs of the complex filter. Considering the image response, then according to (3.6), an inverse version of the signal in Q channel is required. In that case, the signals that are inserted in I and Q channels of the complex filter have equal amplitude but phase 90° and 180°, respectively.

The achieved center frequency is 1 and 2 MHz, while the bandwidth is 0.98 MHz and 1.95 MHz for Bluetooth and ZigBee respectively. Also, the maximum value of in-band group delay variation was simulated equal to 0.79 μs for Bluetooth

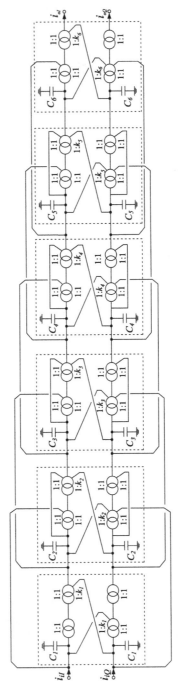

Fig. 3.11 Complete twelfth-order complex filter using current mirrors

Table 3.1 Capacitor values for the passive prototype filter and complex filter in Fig. 3.11

Element of LC filter	Normalized values	Element of complex filter	Bluetooth (pF)	ZigBee (pF)
C_{p1}	82.38 mF	C_1	22.12	11.06
L_{p2}	225.1 mH	C_2	60.42	30.21
C_{p3}	307.5 mF	C_3	82.54	41.27
L_{p4}	307.5 mH	C_4	82.54	41.27
C_{p5}	225.1 mF	C_5	60.42	30.21
L_{p6}	82.38 mH	C_6	22.12	11.06

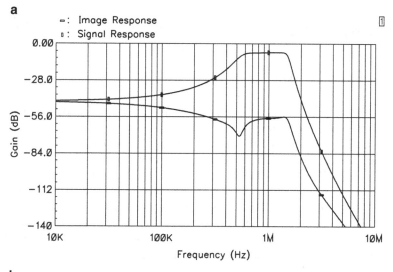

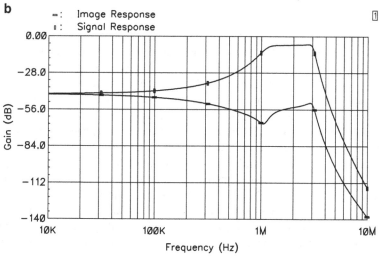

Fig. 3.12 Frequency responses for signal and image sides in (**a**) Bluetooth and (**b**) ZigBee standard

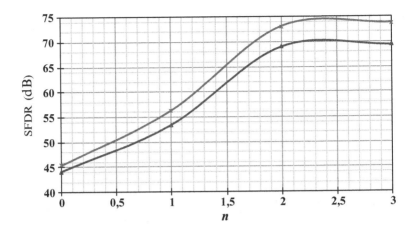

Fig. 3.13 The simulated SFDR versus parameter n

and 0.45 μs for ZigBee. Also, the achieved minimum IRR is 49 dBc for Bluetooth and 44.8 dBc for ZigBee, respectively.

The linear performance of the filter was evaluated through a Periodic State Space (PSS) analysis. The adjacent channels are equally spaced by $\Delta f = 1$ MHz for Bluetooth and $\Delta f = 2$ MHz for ZigBee. Thus, the two tones applied in the filter are located at the frequencies $f_1 = f_{IF} + \Delta f \times n$ MHz and $f_2 = f_{IF} + 2\Delta f \times n$ MHz, where f_{IF} is the center frequency of the Bluetooth/ZigBee filter and $n = f_2 - f_1$.

With regards to the in-band linearity, two closely spaced tones (1, 1.1 MHz for Bluetooth and 2, 2.2 MHz for ZigBee) have been applied at the input of the filter. The simulated IIP3 values for Bluetooth and ZigBee filters were −73.86 and −73.25 dBm, respectively. In the case of Bluetooth filter, the corresponding values of IIP3 for out-of-band near (2 and 3 MHz) and distant (4 and 7 MHz) blockers were −57.5 and −31.35 dBm, respectively. In the case of ZigBee filter, the corresponding values of IIP3 for out-of-band near (4 and 6 MHz) and distant (8 and 14 MHz) blockers were −59.15 and −35.75 dBm, respectively.

The noise was integrated over a 1 MHz range for Bluetooth and 2 MHz for ZigBee and the simulated rms values of the input referred noise (INOISE) were 17.7 and 24.1 nA, respectively. Considering that the $SFDR = (2/3) \cdot (IIP3 - INOISE)$, then a plot of the SFDR versus parameter n is depicted in Fig. 3.13. All the above results are summarized in Table 3.2.

Simulations results concerning the selectivity, IRR and in-band group delay variation verify that the requirements of Bluetooth and ZigBee standards are fulfilled. The evaluation of in-band and out-of-band linearity shows that the filter has attractive characteristics. Taking also into account that current mirrors have the capability of operating in a low voltage environment, they can be considered as very promising active elements for realizing filters in modern low-IF receivers.

Table 3.2 Performance characteristics of the complex filters

Performance factor	Complex filter	
	Bluetooth	ZigBee
Supply voltage (V_{DD})	1.5 V	1.5 V
Power dissipation	3.2 mW	3.2 mW
Center frequency (f_{IF})	1 MHz	2 MHz
Bandwidth	0.54 MHz–1.52 MHz	1.09 MHz–3.05 MHz
In-band group delay variation	0.79 μs	0.45 μs
Input referred noise	17.7 nA	24.1 nA
Image rejection ratio	49 dB	44.8 dB
First blocker attenuation ($f_{IF} + \Delta f$)	34.8 dBc	35.5 dBc
Second blocker attenuation ($f_{IF} + 2\Delta f$)	71.28 dBc	71.97 dBc
Third blocker attenuation ($f_{IF} + 3\Delta f$)	92.7 dBc	93.26 dBc
In-band IIP3	−73.86 dBm	−73.25 dBm
Out-of-band IIP3	−57.5 dBm @ 2&3 MHz	−59.15 dBm @ 4&6 MHz
	−32.45 dBm @ 3&5 MHz	−35.75 dBm @ 6&10 MHz
	−31.35 dBm @ 4&7 MHz	−35.18 dBm @ 8&14 MHz
In-band SFDR	45.42 dB	44.1 dB
Out-of-band SFDR	56.3 dB @ 2&3 MHz	53.4 dB @ 4&6 MHz
	73 dB @ 3&5 MHz	69 dB @ 6&10 MHz
	73.7 dB @ 4&7 MHz	69.4 dB @ 8&14 MHz

3.3 Complex Filter Design Using Topological Emulation of LC Passive Filters

In previous section, the design of high-order Leapfrog complex filters using low-voltage current mirrors was presented. In the current section, the topological emulation of LC ladder filters will be described, which is an alternative method for the synthesis of high-order complex filters. Finally, based on the aforementioned methods, two twelfth-order differential complex filters that fulfill the specifications of Bluetooth standard are fabricated using AMS 0.35 μm C35B4 CMOS process. The performance of the fabricated chip is verified through extensive experimental measurements.

In the case of topological emulation of LC ladder prototypes, there is a one-to-one correspondence between the elements of the active and passive realizations. The inductors are substituted by appropriate active configurations which simulate their operation, while capacitors remain unaffected. Obviously, the substitution of passive resistors is a more simple case. This substitution is done in a way that the current that flows through a passive element is equal to the current that flows through its active equivalent. Thus, the active filter preserves the topology and the physical operation of the passive prototype filter. Consequently, some attractive features of the derived active filters are the low sensitivity passband property and that the optimization of dynamic range can be performed more easily. The complex filter synthesis using the topological emulation of LC ladder prototypes is

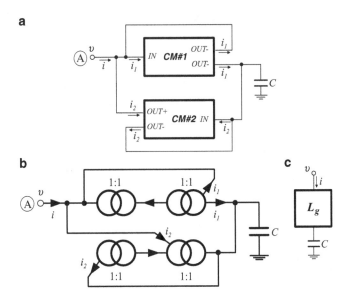

Fig. 3.14 Realization of grounded inductance simulator (**a**) FBD (**b**) topology using current mirrors (**c**) symbol

a three-step procedure: (1) derivation of active subcircuits which simulate the operation of passive inductors (2) transposition of each real active equivalent to the corresponding complex one and (3) synthesis of complex filter topology using the complex subcircuits derived from previous step.

From the above, the first step is the derivation of active configurations using current mirrors that simulate the operation of grounded and floating inductors [16]. The block diagram of the grounded inductance and its corresponding topology using current mirrors is depicted in Fig. 3.14. The current which is entered to the input of CM#1 is given by $i_1 = g_{m1}v$, where g_{m1} is the small-signal transconductance parameter of the input transistor of current mirror CM#1. Due to feedback scheme in CM#2, no current is inserted in the current mirror, thus the current that flows through the capacitor C is equal to i_1. The voltage at the capacitor node is expressed by

$$v_C = \frac{g_{m1}}{Cs}v \qquad (3.22)$$

From (3.22), it is obvious that CM#1 performs the integration of the voltage v at node A, while the current mirror CM#2 converts the voltage v_C to a current $i_2 = g_{m2}v_C$, where g_{m2} is the transconductance parameter of the input transistor of CM#2. Using (3.22), it is derived that

$$i_2 = \frac{g_{m1}g_{m2}}{Cs}v \qquad (3.23)$$

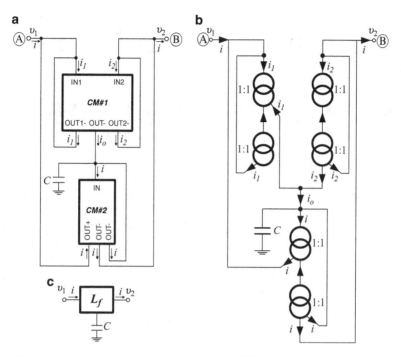

Fig. 3.15 Realization of floating inductance simulator (**a**) FBD (**b**) topology using current mirrors (**c**) symbol

The feedback path in CM#1 establishes that the current i_2 is equal to the current which is entered to the node A, i.e. $i_2 = i$. Thus, the value of the equivalent impedance at node A is expressed by

$$Z_A \equiv \frac{v}{i} = \frac{C}{g_{m1}g_{m2}} s \qquad (3.24)$$

From (3.24), it is obvious that the configuration shown in Fig. 3.14 simulates a grounded inductance with a value

$$L_{eq} = \frac{C}{g_{m1}g_{m2}} \qquad (3.25)$$

In a similar way, the block diagram of a floating inductance simulator using current mirrors is given in Fig. 3.15. The output current of the differential input unity-gain current mirror (CM#1) is $i_o = i_{i1} - i_{i2}$. Due to the fact that no current is inserted in CM#2, then the difference of voltages v_1 and v_2 is integrated at the capacitor node and is given by

$$v_C = \frac{g_{m1}}{Cs}(v_1 - v_2) \qquad (3.26)$$

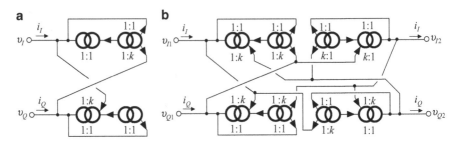

Fig. 3.16 Realization of (**a**) grounded and (**b**) floating negative admittance

As in the previous case, CM#2 converts the capacitor's voltage to a current, while the feedback scheme in CM#1 implies that the currents at nodes A and B are determined by CM#2. By following a routine algebraic analysis, it can be easily verified that the topology shown in Fig. 3.15 simulates a floating inductance, where its value is given by (3.25). It is obvious from the configurations in Figs. 3.14 and 3.15, that only grounded capacitors are needed.

As it has already been mentioned, the transposition of real passive prototype filter into the corresponding complex passive prototype filter is performed by transposing each frequency dependent element into its corresponding complex counterpart. This is achieved by substituting the variable s in the impedance of capacitors and inductors by its shifted version $s - j\omega_{IF}$. More specifically, the impedance of a real capacitor, which is given by $Z(s) = 1/Cs$, is transformed into the complex capacitance

$$Z(s) = \frac{1}{C(s - j\omega_{IF})} \tag{3.27}$$

As a result, each complex capacitor will be constructed from the parallel connection of a real capacitor with value C_r and a complex admittance with value $G = -j\omega_{IF}C_r$. In a similar way, the impedance of a real inductor, i.e. $Z(s) = Ls$, is transformed into the complex inductance

$$Z(s) = L(s - j\omega_{IF}) \tag{3.28}$$

Equation 3.28 reveals that the complex inductor will be formed by a series connection of a real inductor L_r and a complex resistor with a value $R = -j\omega_{IF}L_r$.

The proposed topology of a grounded admittance, realized using current mirrors as active elements, is depicted in Fig. 3.16a, where k is a scaling factor. The expressions of quadrature currents i_I and i_Q are expressed by (3.29a)–(3.29b).

$$i_I = kg_m \cdot v_Q \tag{3.29a}$$

$$i_Q = -kg_m \cdot v \tag{3.29b}$$

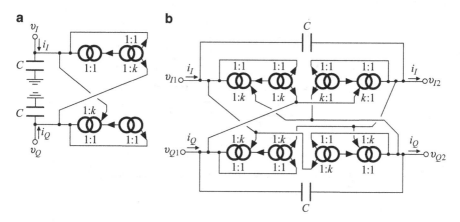

Fig. 3.17 Realization of (a) grounded and (b) floating complex capacitor

Taking into account that the complex voltage (v_c) and current (i_c) are defined as $v_c = v_I + jv_Q$ and $i_c = i_I + ji_Q$, it is readily obtained that the admittance of the topology is given by the expression $G \equiv i_c/v_c = -jkg_m$. Following the same approach, the realization of a floating complex admittance of the same value is given in Fig. 3.16b.

Having available these complex elements, the realization of complex capacitors is performed by connecting in parallel a real capacitor with the corresponding complex admittance, as shown in Fig. 3.17.

Obviously, the realization of complex inductors is performed by connecting in series a real inductor with a complex negative resistor. The concept of an alternative realization that could offer a reduction of the total required active elements is depicted in Fig. 3.18. Based on the topologies presented in Figs. 3.14 and 3.15, then by connecting an admittance G in parallel with the capacitor, the impedance of the derived topologies is given by (3.30)

$$Z = L_{eq} \cdot s + \frac{G}{g_m^2} \tag{3.30}$$

where $L_{eq} = C/g_m^2$. That is, having available the negative grounded admittance with a value $-jkg_m$, shown in Fig. 3.16, the corresponding complex floating and grounded inductor emulators are given in Fig. 3.19. Setting $G = -jkg_m$ in (3.30), it is easily derived that the expression of impedance for both topologies is $Z = L_{eq}s - j$ (k/g_m). Thus, the values of capacitors and scaling factors are calculated by (3.20b) and (3.21b), respectively.

The real passive elements as well as their complex active equivalents are summarized in Table 3.3. The last step is the design of the high-order complex filter using the topological method. As depicted in Fig. 3.20, at first the passive elements of the real passive prototype filter are replaced by the corresponding

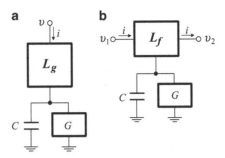

Fig. 3.18 Realization of a real synthetic (**a**) grounded and (**b**) floating inductor

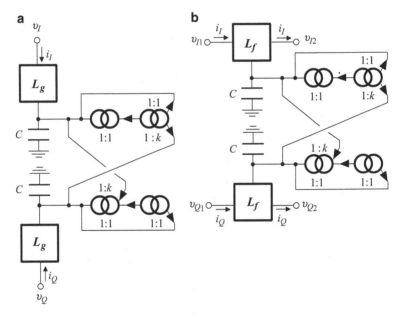

Fig. 3.19 Realization of a complex (**a**) grounded and (**b**) floating inductor

complex passive elements. The next step is the synthesis of complex filter topology using the appropriate complex subcircuits of Table 3.3. Finally, concerning the passive resistors, each resistor at the inputs of the filter is replaced by a diode connected transistor, while each termination resistor is realized by a current mirror in order the output currents to be separately available.

Table 3.3 Topologies of complex elements using current mirrors

Passive element	Topology of complex equivalent

3.4 Experimental Results of twelfth-Order Complex Filters

The performance of the proposed building blocks will be evaluated through two complex filter design examples, where their center frequency is 2 MHz and will meet the requirements of Bluetooth standard [17]. Additionally, the complex filters have been designed in a differential form in order to be compliant with modern RF systems. The following Section discusses the employed general measurement setup, the appropriate interfaces at the inputs and outputs of the filters for the V-I and I-V conversion, as well as the PCB for the experimental verification of the fabricated chip.

According to Sect. 3.2.2, a complex filter that is based on a sixth-order Butterworth lowpass filter with 500 kHz cutoff frequency, will comply adequately to the specifications of Bluetooth standard. The complex filter derived by using the Leapfrog method has been already shown in Fig. 3.11. The second complex filter topology will be derived by the topological method presented in Fig. 3.20, where the order of filter is $n = 6$. The values of capacitors and scaling factors in both realizations are calculated by (3.20a)–(3.20b) and (3.21a)–(3.21b), respectively.

The low-voltage current mirror which is also used in previous chapters have been employed with a dc bias current $I_0 = 12\ \mu\varsigma$ and supply voltages $V_{DD} = 1.5$ V and $V_{DC} = 1.2$ V. Considering the MOS transistors parameters provided by the AMS C35B4 process, the values of NMOS transistor aspect ratios are 3 μm/1 μm for M_{ni} (i = 1,2,...) and 7 μm/0.6 μm for M_{n2i} (i = 1,2,...). The PMOS transistors have an aspect ratio 15 μm/2 μm. Under the above bias conditions, the input resistance of the current mirrors is $R_{in} = 1/g_{m,Mn11} = 11.4$ kΩ.

The values of capacitors for the passive prototype and the complex filters are given in Table 3.4. It must be noted at this point, that due to limited die area and taking advantage of the fact that the same capacitor values are used in both filters, then by using an appropriate switching scheme, the two configurations can share the same capacitors. Finally, the reference current for the biasing of the complex filters is generated by the topology given in Fig. 3.21, where the corresponding values of the NMOS and PMOS transistor aspect ratios as well as the other component values are given in Table 3.5.

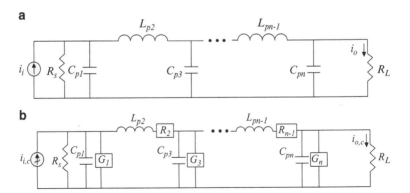

Fig. 3.20 Synthesis of complex filters using topological method (**a**) passive prototype of an nth-order real lowpass filter (**b**) passive prototype of an nth-order complex filter and (**c**) topology of nth-order complex bandpass filter using current mirrors

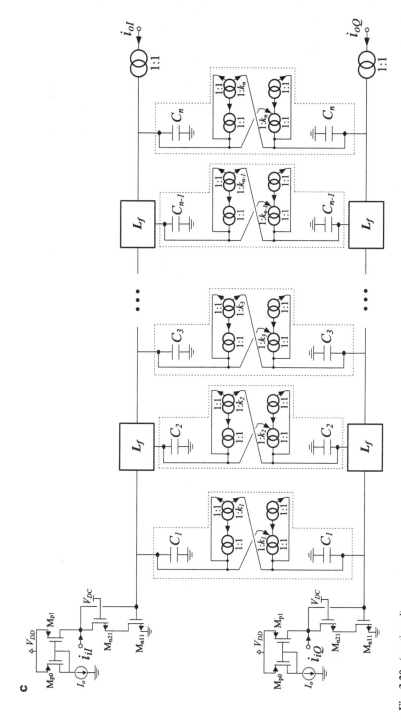

Fig. 3.20 (continued)

Table 3.4 Capacitor values for the passive prototype filter and complex filters

Element of LC filter	Normalized values	Element of complex filter	Leapfrog/ topological (pF)
C_{p1}	82.38 mF	C_1	14.45
L_{p2}	225.1 mH	C_2	39.49
C_{p3}	307.5 mF	C_3	53.94
L_{p4}	307.5 mH	C_4	53.94
C_{p5}	225.1 mF	C_5	39.49
L_{p6}	82.38 mH	C_6	14.45

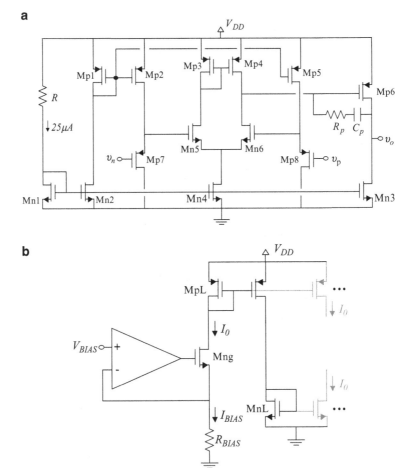

Fig. 3.21 (a) Topology for the generation of bias current and (b) the circuit of the utilized two-stage amplifier

Table 3.5 Aspect ratios of the MOS transistors and component values of the amplifier in Fig. 3.21

Transistor	W/L (μm/μm)
M_{n1}–M_{n3}	20/1
M_{n4}	40/1
M_{n5}, M_{n6}	25/1
M_{p1}–M_{p6}	40/1
M_{p7}–M_{p8}	50/0.5
M_{ng}	10/1
M_{pL}	25/1
M_{nL}	15/1
Component	Value
C_p	4 pF
R_p	2.2 kΩ
R	34 kΩ
R_{BIAS}	7.62 kΩ

3.4.1 Test Setup and Measurements

A microphotograph of the fabricated chip is depicted in Fig. 3.22, where it should be mentioned that the two complex filters have been laid out very carefully, since the mismatches in current mirrors and capacitors directly translate to the degradation of overall performance, such as the reduction of IRR and frequency response of the filters. The fabricated chip was tested in a PCB that was implemented through ORCAD PCB Editor 16.0 of Cadence, where its photograph is depicted in Fig. 3.23. Extended measurements were taken using the HP4395A Network/Spectrum analyzer. The employed general measurement setup is demonstrated in Fig. 3.24.

The required quadratic differential input currents are produced by employing the hybrid 90° device QE-19-4442 component provided by the Pulsar Microwave Corporation, in combination with THS4130 (Texas Instruments) for the single to differential conversion and AD844 (Analog Devices) components configured as V-I converters. THS4130 devices are also used at the output of the filters for the differential I-V conversion. Concerning the quadrature signal generation, polyphase RC network can be used for this procedure [18]. However, this type of filters can only generate balanced quadrature signals for a narrow band and thus, cannot be used to measure the attenuation and image rejection. Furthermore, process variations and component mismatching affect the RC time constants, which will lead to unbalanced quadrature signals. For these reasons, the quadrature signal generation was implemented through the hybrid 90° device.

The simulated and measured frequency responses of both filters are demonstrated in Fig. 3.25, where a small deviation between them at high frequencies can be noticed due to the parasitics inserted by external devices. The frequency responses for the signal and image sides are given in Fig. 3.26, where the IRR is 28.2 dBc for the Leapfrog filter, and 27.9 dBc for the topological filter. The corresponding Bluetooth specification is 20–30 dBc.

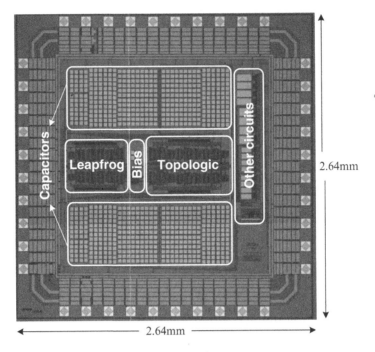

Fig. 3.22 Die microphotograph of the fabricated chip

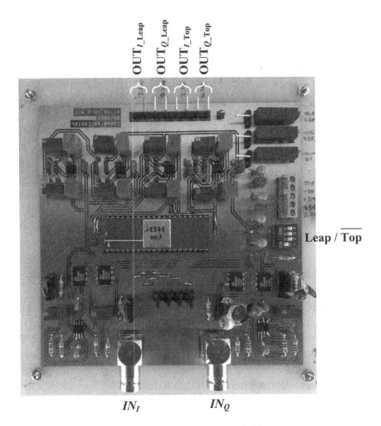

Fig. 3.23 Photograph of the PCB used to test the fabricated chip

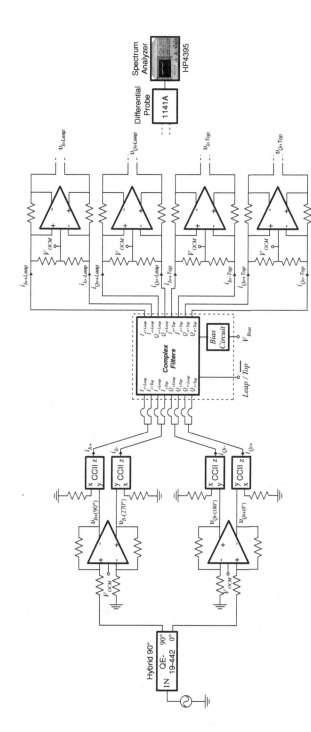

Fig. 3.24 General measurement setup for the experimental verification of twelfth-order Butterworth complex filters

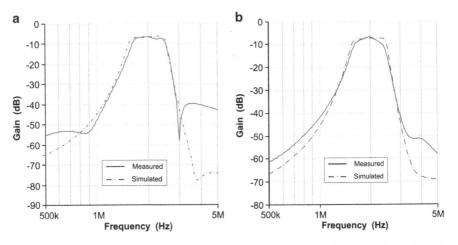

Fig. 3.25 Simulated and measured frequency responses of (**a**) leapfrog and (**b**) topological complex filter

Moreover, the first-blocker attenuation of the Leapfrog and topological complex filter are 56.9 and 36.2 dBc, respectively. Thus, the two filters over satisfy the Bluetooth requirement of 11 dBc.

The measured group delay responses are given in Fig. 3.27, where the maximum variation is 0.56 μs for the Leapfrog filter and 0.41 μs for the topological filter. According to Bluetooth specifications, the in-band group delay variation should be less than 1 μs to eliminate intersymbol interference.

The linear performance of the complex filters has been evaluated by performing a two-tone test. For this purpose, two voltages with frequencies 1.95 and 2.05 MHz and variable amplitude have been applied at the input of each filter and they are added using the Agilent 11636 C Agilent Power Combiner/Divider. The derived spectrum plots at 1% (−40 dB) level of third-order Intermodulation Distortion (IM3) are given in Fig. 3.28 and correspond to an input power (P_{in}) −4.89 dBm for the Leapfrog filter and −5.39 dBm for the topological filter. Thus, the third-order Input referred Intercept Points (IIP3) will be 15.11 and 14.6 dBm, respectively. The noise has been integrated within a 1 MHz range and the derived rms values of the input referred noise (INOISE) were −45.43 dBm for the Leapfrog filter and −42.6 dBm for the topological filter. Recalling (2.13) for the calculation of SFDR, i.e. SFDR = (2/3)·(IIP3 − INOISE), the calculated value for the in-band SFDR was 40.4 dB for the Leapfrog filter. Also, by applying two tones with frequencies 3 and 4 MHz, the calculated value of the out-of-band SFDR was 53.8 dB. The corresponding values of SFDR for the topological filter are 39.2 and 52.9 dB, respectively. All the experimental results are summarized in Table 3.6. According to this Table, it can be concluded that both filters fulfill the requirements of Bluetooth standard. Additionally, the Leapfrog complex filters offers reduced power dissipation compared with the topological complex filter.

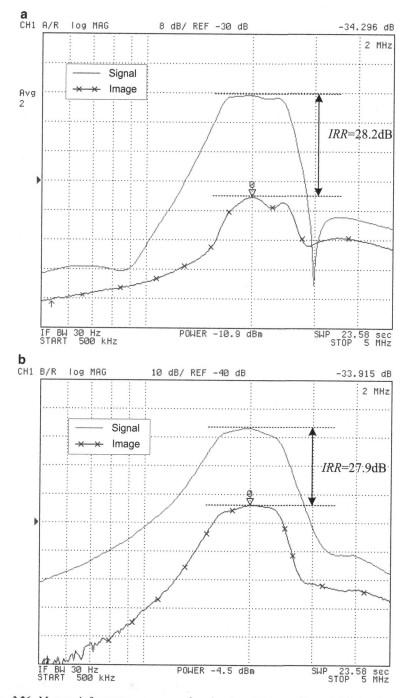

Fig. 3.26 Measured frequency responses for signal and image sides of (**a**) leapfrog and (**b**) topological complex filter

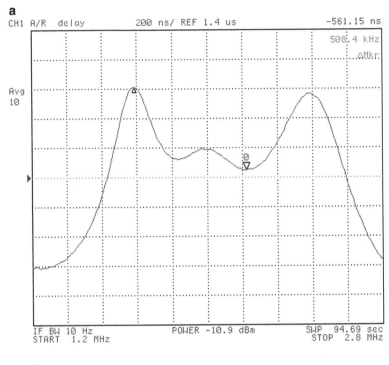

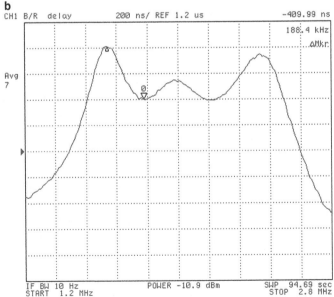

Fig. 3.27 Measured group delay variation of (**a**) leapfrog and (**b**) topological complex filter

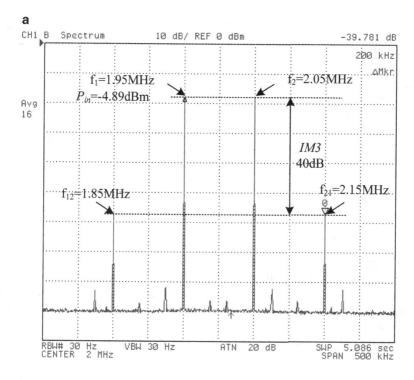

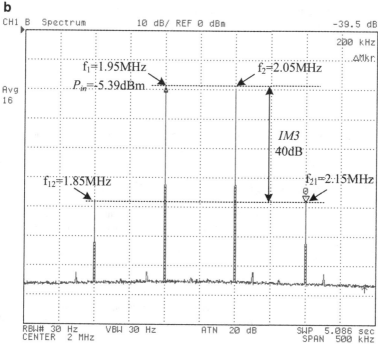

Fig. 3.28 Measured linear performance of (**a**) leapfrog and (**b**) topological complex filter

Table 3.6 Experimental results of the twelfth-order complex filters

Performance factor	Leapfrog	Topological
Supply voltage (V_{DD})	1.5 V	1.5 V
Power dissipation	4.71 mW	7.2 mW
Center frequency (f_{IF})	2 MHz	2 MHz
Bandwidth	1.6 MHz–2.5 MHz	1.62 MHz–2.4 MHz
Group delay variation	0.56 μs	0.41 μs
INOISE (rms)	−45.3 dBm	−42.6 dBm
IRR	28.2 dB	27.9 dB
First-blocker attenuation	56.9 dBc	36.2 dBc
In-band SFDR	40.4 dB	39.2 dB
Out-of-band SFDR	53.8 dB	52.9 dB

3.5   Summary

This chapter presented novel complex resistorless filter topologies with grounded capacitors, realized using current mirrors as active elements. At first, a twelfth-order complex Leapfrog filter has been designed in order to meet the Bluetooth and ZigBee standard requirements. As a second design example, two twelfth-order differential complex filters using the Leapfrog and topological emulation of LC ladder filter methods have been fabricated in AMS 0.35 μm CMOS process. The obtained experimental results verified the correct operation of both filters. Some essential results are that the Leapfrog filter has IRR = 28.2 dBc, in-band group delay variation equal to 0.56 μs and SFDR = 40.4 dB, while the corresponding values for topological filter are 27.9 dBc, 0.41 μs and 39.2 dB.

References

1. W. M. Snelgrove, and A. S. Sedra, "State-space synthesis of complex analog filters", in *Proc. European Conference on Circuit Theory and Design (ECCTD)*, pp. 420–424, Aug. 1981.
2. G. R. Lang, and P. O. Brackett, "Complex analogue filters", in *Proc. European Conference on Circuit Theory and Design (ECCTD)*, pp. 412–419, Aug. 1981.
3. K. Martin, "Complex signal processing is not complex", *IEEE Transactions on Circuits and Systems I*, vol. 51, no.9, pp. 1823–1836, Sept. 2004.
4. A. Worapishet, R. Sithikorn, A. Spencer, and J. B. Hughes, "A multirate switched-current filter using class-AB cascaded memory", *IEEE Transactions on Circuits and Systems II*, vol.53, no.11, pp. 1323–1327, Nov. 2006.
5. C. N. Y. Voo, M. Teplechuk, and J. I. Sewell, "General synthesis of complex analogue filters", *IEE Proc., Part-G*, vol. 152, no.1, pp. 7–15, Feb. 2005.
6. C. Psychalinos, "Low-voltage log-domain complex filters", *IEEE Transactions on Circuits and Systems I*, vol.55, no.11, pp. 3404–3412, Dec. 2008.
7. P. Andreani, and S. Mattisson, "On the use of Nauta's transconductor in low-frequency CMOS g_m-C bandpass filters", *IEEE Journal of Solid-State Circuits*, vol. 37, no.2, pp.114–124, Feb. 2002.

8. J. B. Hughes, A. Spencer, A. Worapishet, and R. Sithikorn, "1 mW CMOS polyphase channel filter for Bluetooth", *IEE Proc. Circuits Devices and Systems*, vol. 149, no.5/6, pp. 348–354, Oct./Dec. 2002.

9. A. A. Emira, and E. Sanchez-Sinencio, "A pseudo differential complex filter for Bluetooth with frequency tuning", *IEEE Transactions on Circuits and Systems II*, vol. 50, no.10, pp. 742–754, Oct. 2003.

10. B. Guthrie, J. Hughes, T. Sayers, and A. Spencer, "A CMOS Gyrator Low-IF filter for a dual-mode Bluetooth/ZigBee transceiver", *IEEE Journal of Solid-State Circuits*, vol. 40, no.9, pp.1872–1879, Sept. 2005.

11. X. Zhang, X. Ni, M. Iwahashi, and N. Kambayashi, "Realization of universal active complex filters using CCIIs and CFCCIIs", *Analog Integrated Circuits and Signal Processing.*, vol. 20, no.1, pp. 129–137, Feb.1999.

12. M. T. Abuelma'atti, and S. M. Al-Shahrani, "A new polyphase current-mode filter using programmable-gain current-controlled current conveyor", *WSEAS Transactions on Electronics*, vol. 4, no.2, pp. 138–141, Oct. 2005.

13. H. Alzaher, "A fully integrated MOS-C current-mode IF filter for Bluetooth", in *Proc. IEEE International Conference on Electronics Circuits and Systems (ICECS)*, pp. 579–582, Shariqah, United Arab Emirates, Dec. 2003.

14. B. Razavi, RF Microelectronics, Englewood Cliffs, NJ: Prentice Hall, 1997.

15. C. Laoudias, C. Psychalinos, "Low-Voltage Bluetooth/ZigBee Complex Filter Using Current Mirrors", in *Proc. IEEE International Symposium on Circuits and Systems (ISCAS)*, Paris, pp. 1268–1271, May 2010.

16. C. Psychalinos, and A. Spanidou, "Current amplifier based grounded and floating inductance simulators", *International Journal of Electronics and Communications (AEUE)*, vol.60, no.2, pp.168–171, Feb. 2006.

17. C. Laoudias, C. Psychalinos, "1.5 V Complex Filters using current mirrors", *IEEE Transactions on Circuits and Systems II*. DOI: 10.1109/TCSII.2011.2161169.

18. P. Andreani, S. Mattisson, and B. Essink, "A CMOS g_m-C polyphase filters with image band rejection", in *Proc. 26th European Solid-State Circuits Conference*, pp. 244–247, Sept. 2000.

Chapter 4
Filters for Biomedical Applications

Abstract This chapter focuses on the design of novel topologies that are suitable for realizing wavelet filter functions. These filters are extensively used in several biomedical applications and especially in low-voltage/low-power implantable devices for the detection and analysis of cardiac signals. As in previous chapters, low-voltage current mirrors will be employed as active elements, providing, thus, the advantages of resistorless topologies and the electronic adjustment capability of their frequency characteristics. In addition, by using MOS transistors biased in subthreshold region, there is the benefit of ultra low-voltage (0.5 V) operation. The efficiency of the proposed filters is verified through simulation results by employing TSMC 130 nm CMOS process, where the most important performance factors are considered.

Keywords Analog wavelet filters • Biomedical applications • Current mirror filters • Current mode circuits • Implantable pacemakers

4.1 Introduction

Cardiac signal detection and analysis is one of the primary functions of an implantable pacemaker. Their main operation is the analysis of electrocardiogram (ECG) signal, which is the recording of potential differences between areas positively and negatively charged. The potential differences arise in the outer membrane surface of myocardial fibers during depolarization and repolarization of the myocardium. Thus, the ECG represents the electrical events associated with cardiac stimulation and provides information on the cardiac chamber enlargement, heart rate, rhythm, and intracardiac conduction. All these are composing the ECG signal, which consists of P, Q, R, S, T and U waves, as depicted in Fig. 4.1.

Thus, an ECG signal in a normal cardiac pulse is constructed by a P wave, a QRS complex and a T wave. A small U wave may also be visible in 50–75% of ECGs but the most times is neglected. The P wave is caused by the repolarization of atria,

C. Laoudias and C. Psychalinos, *Integrated Filters for Short Range Wireless*
and Biomedical Applications, SpringerBriefs in Electrical and Computer Engineering,
DOI 10.1007/978-1-4614-0260-2_4, © Springer Science+Business Media, LLC 2012

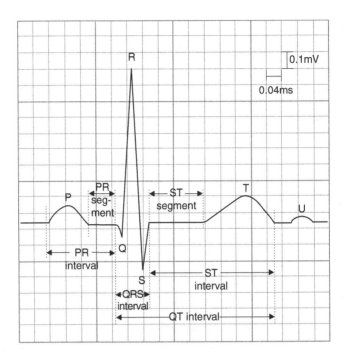

Fig. 4.1 A standard ECG signal with characteristic P, Q, R, S, T, U points

while the QRS complex corresponds to the current that causes contraction of the left and right ventricles. The T wave represents the repolarization of the ventricles. These waves as well as the QT and PR intervals are meaningful parameters in the detection of heart's abnormalities. Differences between normal ECGs and those derived from a patient can be found by examining their length, frequency and amplitude. Therefore, the objective to accurately analyze an ECG signal is very important within the framework of diagnosis.

However, the feature extraction from ECG signals is a very complicated procedure, since they usually share the same spectral band with noise components and interference. For this purpose, appropriate mathematical tools must be employed in order to investigate the location of the most interested characteristic points and thus, to discriminate normal from abnormal cardiac patterns. In the literature, several approaches for cardiac signal analysis have been reported, where the most of them are algorithms implemented by artificial neural networks, filterbanks, non-linear transformations and the wavelet transform. The wavelet based signal processing methods, though relatively new, demonstrate the potential for accurate feature extraction from noisy cardiac signals. The wavelet transform is a computational complex method and requires a lot of processing power. There are two subclasses of WT: the Discrete Wavelet Transform (DWT) and the Continuous Wavelet Transform (CWT). Due to the ultra low-power requirements of implantable biomedical devices, the realization of wavelet transform digitally is not suitable because of the employment of A/D converters. This is not the case for the implementation of continuous wavelet transform in the analog part of a system, as this can be done in low-power and the computational delay will be minimized. However, by using analog circuits it is not possible to compute the wavelet transform with great accuracy. Hence, the use of approximation methods for the realization of continuous wavelet transform using analog circuits is unavoidable [1].

A number of analog wavelet filters for detecting the QRS complexes of ECGs have been already introduced in the literature [2–7]. The topologies in [2–6] have been realized by employing the concept of log-domain filtering. For this purpose, non-linear transconductor cells constructed from bipolar transistors have been employed as active elements. The topology in [7] is realized by employing Operational Transconductance Amplifiers (OTAs), constructed from MOS transistors in triode region, as active blocks. All the aforementioned realizations have been performed by employing a bias voltage greater than or equal to 1 V.

As has been already mentioned in previous chapters, current mirrors are very simple and reliable building blocks for performing current-mode analog signal processing. This stems from their very simple structure compared with that of other elements such as Operational Amplifiers. They have been successfully employed for realizing analog filter transfer functions where an electronic adjustment of the cutoff frequency is offered. This is originated from the small-signal nature of the current mirrors' input resistance which is equal to $1/g_m$, where g_m is the small signal transconductance parameter of an MOS transistor in strong inversion. This parameter can be electronically controlled through the corresponding dc bias current of the current mirror [8, 9].

In this chapter, novel wavelet filter topologies that are suitable for detecting the QRS complex of ECGs are introduced. Due to the fact that they are constructed from current mirrors operating in subthreshold region, they offer the benefit of ultra low-voltage (0.5 V) operation. Thus, at first the Functional Block Diagrams (FBDs) of the wavelet filter topologies are presented followed by the basic building blocks for realizing these FBDs. Finally, the efficiency of the proposed filters is verified through simulation results by employing TSMC 130 nm CMOS process, where the most important performance factors are considered.

4.2 Analog Wavelet Filter Topologies

Let us consider the arbitrary order analog wavelet filter function given by (4.1), where A_i, $(i = 0,\ldots,n)$, B_j, $(j = 0,\ldots,n-1)$ are real coefficients and $B_j > 0$.

$$H(s) = \frac{A_n s^n + A_{n-1} s^{n-1} + \ldots + A_1 s + A_0}{s^n + B_{n-1} s^{n-1} + \ldots + B_1 s + B_0} \tag{4.1}$$

The FBDs of possible candidates for realizing the transfer function in (1) are given in Fig. 4.2. The FBD in Fig. 4.2a corresponds to a Follow the Leader

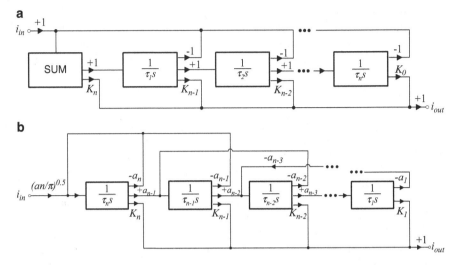

Fig. 4.2 FBD of a current-mode filter derived from (**a**) FLF method and (**b**) emulation of an orthonormal ladder filter

Feedback (FLF) topology [10], while the FBD in Fig. 4.2b to the operational emulation of an orthonormal filter [11]. The selection of these two filter design methods was made by considering the most important requirements in the design of high-order filters, namely the sensitivity to active and passive component mismatch, dynamic range, noise and sparsity.

The transfer function that is derived according to the FBD in Fig. 4.2a is given by

$$H(s) = \frac{K_n s^n + \frac{K_{n-1}}{\tau_1} s^{n-1} + \frac{K_{n-2}}{\tau_1 \tau_2} s^{n-2} \ldots + \frac{K_1}{\tau_1 \tau_2 \ldots \tau_{n-1}} s + \frac{K_0}{\tau_1 \tau_2 \ldots \tau_n}}{s^n + \frac{1}{\tau_1} s^{n-1} + \frac{1}{\tau_1 \tau_2} s^{n-2} \ldots + \frac{1}{\tau_1 \tau_2 \ldots \tau_{n-1}} s + \frac{1}{\tau_1 \tau_2 \ldots \tau_n}} \tag{4.2}$$

That is, by equating both numerator and denominator coefficients of (4.1) and (4.2), the values of the required time-constants τ_i and scaling factors K_i are readily obtained.

The values of scaling factors in the FBD of Fig. 4.2b are derived through a more complicated process. More specifically, for a given transfer function an appropriate orthonormal ladder prototype filter must be considered in order to realize the poles of the filter function. The values of factors a_i are determined from the values of passive elements of the ladder filter. The values of the scaling factors K_i are calculated by equating the numerator coefficients of the desired transfer function with the corresponding coefficients derived through a linear combination of the states of the orthonormal filter [11]. From the above FBDs, it is obvious that required operations of integration, scaling and summation are easily realized by employing current mirrors as active elements.

4.3 Current Mirrors in Subthreshold Region

A typical topology of a multiple output current mirror, as that presented in previous chapters, with arbitrary gain at each output, is depicted in Fig. 4.3. The required scaling factors are realized by properly sizing the corresponding PMOS and NMOS transistors of the branch associated with this factor. The input resistance of current mirror is equal to $1/g_m$, where g_m is the transconductance parameter of a MOS transistor, and will be used for realizing resistorless filters. Due to the fact that the cutoff frequency of the wavelet filters used for detecting the QRS complex of ECGs is in the order of Hz, this parameter must be realized in the order of nA(pA)/V. Thus, a possible solution is the employment of MOS transistors operated in subthreshold region. For this purpose the gate-source of each transistor must fulfill the condition: $V_{GS}\text{-}V_{th} \leq -100\,\text{mV}$, where V_{th} is the threshold voltage. In this mode of operation the transistor saturates for $V_{DS} \geq 4V_T$, where V_T is the well-known thermal voltage ($\cong 26\,\text{mV}$ at room temperature). Inspecting the topology of Fig. 4.3, it is obvious that a minimum power supply voltage equals to $12V_T$ ($\cong 310\,\text{mV}$) is required in order the transistors to be saturated within the subthreshold region. Thus, a reasonable choice of the supply voltage could be 0.5 V.

The expression of g_m under this condition of operation is given by (4.3) as

$$g_m = \frac{I_0}{nV_T} \tag{4.3}$$

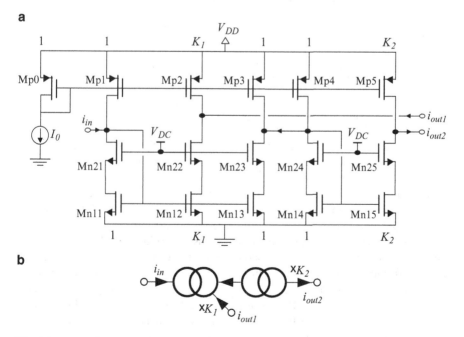

Fig. 4.3 Multiple output current mirror (**a**) topology (**b**) symbol

Fig. 4.4 Lossless integrator
using current mirrors

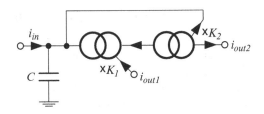

where V_T is the thermal voltage ($\cong$26 mV at 27°C) and n is the subthreshold slope factor.

The realization of a lossless integrator is performed by configuring a current mirror as it is shown in Fig. 4.4. Each output current is described by (4.4)

$$\frac{i_{out}}{i_{in}} = \frac{K_i}{\tau s}, (i = 1, 2) \tag{4.4}$$

where the variable g_m is defined by (4.3). Thus, according to (4.3), the time constant of the integrator, that is given by the formula $\tau = C/g_m$, can be electronically adjusted through the dc current I_o. This is a very attractive feature for deriving different scales of impulse response of the wavelet filter without affecting the topology of the filter.

4.4 Wavelet Filter Design Examples

Let us consider the following fifth-order Gaussian wavelet (scale a_1) filter function [6]

$$H(s) = \frac{8.59s^4 + 1.47 \cdot 10^5 s^3 - 9.69 \cdot 10^7 s^2 + 3.04 \cdot 10^{11} s}{s^5 + 1.28 \cdot 10^3 s^4 + 1.65 \cdot 10^6 s^3 + 9.53 \cdot 10^8 s^2 + 3.86 \cdot 10^{11} s + 6.06 \cdot 10^{13}} \tag{4.5}$$

The corresponding active filter topologies, derived according to the FBDs in Fig. 4.2, are demonstrated in Fig. 4.5 [12].

The behavior of the wavelet filters will be evaluated through simulation results, by employing the Virtuoso Analog Design Environment of the Cadence software. For this purpose Level 49 MOS transistor models provided by TSMC 130 nm CMOS process have been used in simulations. Considering a dc bias current $I_o = 1.2$ nA and supply voltages $V_{DD} = 0.5$ V and $V_{DC} = 240$ mV, the values of the NMOS transistors aspect ratios were 2 μm/20 μm for M_{n1i} and M_{n2i} (i = 1,2,...). The PMOS transistors have an aspect ratio 2 μm/4 μm. Under the above bias conditions, the transconductance of the input transistor of current mirrors is equal to 29.21 nA/V. Thus, the input resistance of current mirrors is equal to $R_{in} = 1/g_{m,Mn11} = 34.23$ MΩ. The calculated values of capacitors for each filter are given in Fig. 4.5. The required scaling factors are realized by properly adjusting the aspect ratio of the corresponding MOS transistors. In the case that a very small value of scaling factor is required (i.e. factors K_2-K_4 in Fig. 4.5a and K_5

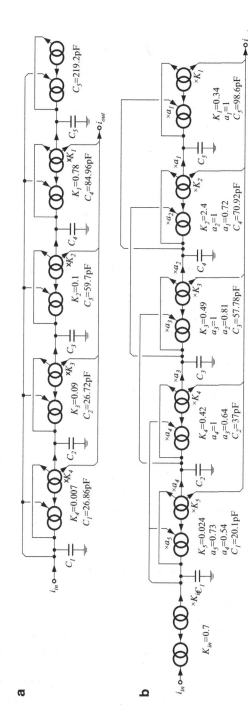

Fig. 4.5 Fifth-order wavelet filter derived from (**a**) FLF method and (**b**) orthonormal prototype filter

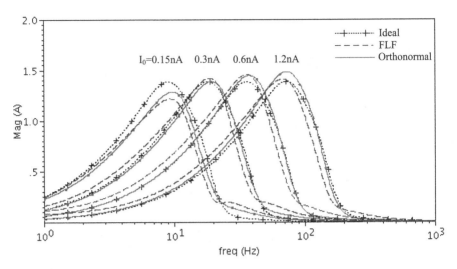

Fig. 4.6 Frequency responses of the wavelet filters in Fig. 4.4 and ideal at scales $a_1(I_o = 1.2\,nA)$, $a_2(I_o = 0.6\,nA)$, $a_3(I_o = 0.3\,nA)$, $a_4(I_o = 0.15\,nA)$

in Fig. 4.5b), this is realized by cascading current mirrors with relative gains much greater than that of the desired total gain.

The dc power dissipation of the filter in Fig. 4.5a was 39.6 nW, while for the filter in Fig. 4.5b was 31.8 nW. Also, the total required capacitance was 417.44 pF for the filter in Fig. 4.5a and 284.4 pF for the filter in Fig. 4.5b. The simulated frequency responses of each one of the considered filters, at different dyadic scales, are given in Fig. 4.6. According to these plots it is concluded that they can be electronically tuned. The observed deviations from those theoretically predicted are caused by the imperfections of current mirrors.

The approximated Gaussian wavelet base must be derived from the impulses responses of the bandpass filters in Fig. 4.5. This is verified in Fig. 4.7, where their electronic tuning capability has been also demonstrated. This is a desired feature from the realization point of view, due to the fact that there is the capability to electronically scale and shift in time and consequently in frequency, the realized Gaussian impulse response.

The linear performance of the filters has been evaluated by employing the Periodic State Space (PSS) analysis tool of the Analog design Environment. The derived plots are given in Fig. 4.8. The levels of the input signal that results to a 1-dB compression point at the outputs of the filters in Fig. 4.5a, b were -177.4 and -166 dBm, respectively.

The noise has been integrated within the passband of each filter and the obtained rms value of the input referred noise was 0.64 pA for the filter in Fig. 4.5a and 1.39 pA for the filter in Fig. 4.5b. Thus, the predicted values of the Dynamic Range at 1-dB compression point will be 53.45 and 58.1 dB, respectively.

Finally, Monte Carlo analysis was also carried out to verify the sensitivities of the filters in Fig. 4.5a, b. The impulse responses depicted in Fig. 4.9 confirm the

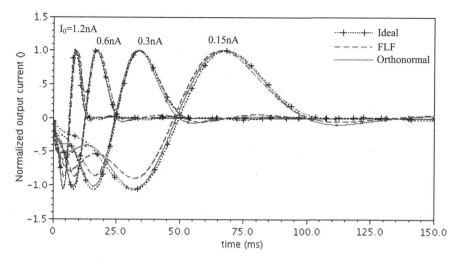

Fig. 4.7 Impulse responses of the wavelet filters in Fig. 4.5 and ideal filter at scales $a_1(I_o = 1.2$ nA), $a_2(I_o = 0.6$ nA), $a_3(I_o = 0.3$ nA), $a_4(I_o = 0.15$ nA)

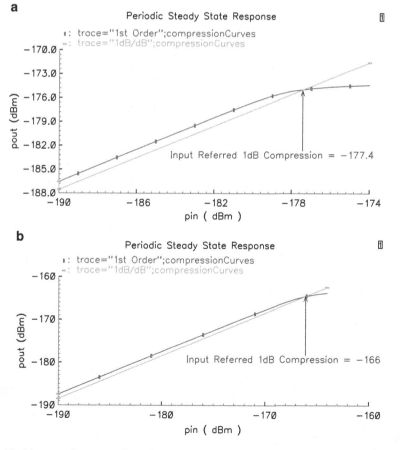

Fig. 4.8 Linear performance of the filter derived from (**a**) FLF method and (**b**) orthonormal prototype filter

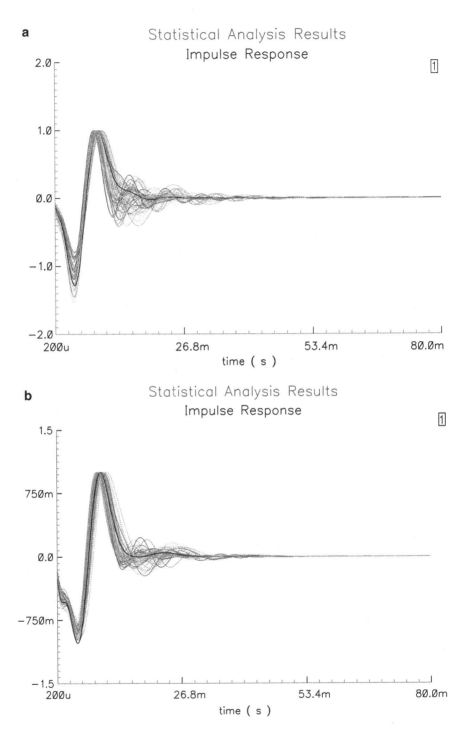

Fig. 4.9 Monte Carlo analysis of the filter derived from (**a**) FLF method and (**b**) orthonormal prototype filter

Table 4.1 Comparison results of wavelet filters

Performance factor	FLF	Orthonormal
Supply voltage (V_{DD})	0.5 V	0.5 V
Bias current (I_0)	1.2 nA	1.2 nA
Power dissipation	39.6 nW	31.8 nW
Total capacitance	417.44 pF	284.4 pF
1-dB input referred compression point	−177.4 dBm	−166 dBm
Input referred noise	0.64 pA	1.39 pA
Dynamic range (DR)	53.45 dB	58.1 dB

reasonable sensitivity characteristics of both filters in process and mismatch variations. It is apparent that the impulse responses are slightly affected, though the original shape is preserved.

The simulation results are summarized in Table 4.1, where it can be concluded that the topology derived from the orthonormal prototype filters has better performance in terms of power, capacitor area, linearity and dynamic range, than that achieved by the FLF topology.

4.5 Summary

Ultra low-voltage wavelet filters are presented in this chapter, where the active blocks are current mirrors operating in subthreshold region. These topologies can be employed in several biomedical applications, like the QRS complex detection from cardiac signals in implantable pacemakers. The analysis and processing of this type of signals is performed by the continuous wavelet transform, which is realized by the proposed analog wavelet filters. The well-known FLF method and the emulation of an orthonormal ladder filter have been utilized for the synthesis of two fifth-order wavelet filter topologies. From the comparison results, it has been concluded that the orthonormal wavelet filter has better performance than the FLF wavelet filter in all terms under consideration.

References

1. S. Haddad, and W. Serdijn, "Ultra low-power biomedical signal processing: an analog wavelet filter approach for pacemakers", Springer Science + Business Media, 2009.
2. S. Haddad, and W. Serdijn, "Mapping the wavelet transform onto silicon: the dynamic translinear approach", in *Proc. IEEE International Symposium on Circuits and Systems (ISCAS)*, Arizona, USA, pp. 621–624, May 2002.
3. S. Haddad, R. Houben, and W. Serdijn, "Analog wavelet transform employing dynamic translinear circuits for cardiac signal characterization," in *Proc. IEEE International Symposium on Circuits and Systems (ISCAS)*, Bangkok, Thailand, pp. 121–124, May 2003.

4. S. Haddad, N. Verwaal, R. Houben, and W. Serdijn, "Optimized dynamic translinear implementation of the Gaussian wavelet transform", in *Proc. IEEE International Symposium on Circuits and Systems (ISCAS)*, pp. 145–148, Vancouver, Canada, May 2004.
5. S. Haddad, S. Bagga, and W. Serdijn, "Log-Domain Wavelet Bases", *IEEE Transactions on Circuits and Systems I*, vol. 52, pp. 2023–2032, Oct. 2005.
6. L. Hongmin, H. Yigang, and Y. Sun, "Detection of Cardiac Signal Characteristic Point Using Log-Domain Wavelet Transform Circuits", *Circuits Systems and Signal Processing*, vol. 27, pp. 683–698, Oct. 2008.
7. P. Agostinho, S. Haddad, J. De Lima, and W. Serdijn, "An ultra low power CMOS pA/V transconductor and its application to wavelet filters", *Analog Integrated Circuits and Signal Processing*, vol. 57, pp. 19–27, Nov. 2008.
8. J. Ramirez-Angulo, M. Robinson, and E. Sanchez-Sinencio, "Current-mode continuous-time filters: two design approaches", *IEEE Transactions on Circuits and Systems II*, vol. 39, pp. 337–341, Jun. 1992.
9. C. Laoudias, and C. Psychalinos, "Universal biquad filters using low-voltage current mirrors", *Analog Integrated Circuits and Signal Processing*, vol. 65, no. 1, pp. 77–88, Oct. 2010.
10. T. Deliyannis, Y. Sun, and J. Fidler, Continuous-time active filter design, CRC Press LLC, 1999.
11. D. Jones, W. Snelgrove, and A. Sedra, "Orthonormal ladder filters", *IEEE Transactions on Circuits and Systems*, vol. 36, pp. 337–343, Mar. 1989.
12. C. Laoudias, C. Beis, and C. Psychalinos, "0.5 V Wavelet Filters Using Current Mirrors", in *Proc. IEEE International Symposium on Circuits and Systems (ISCAS)*, Rio de Janeiro, pp. 1443–1446, May 2011.

Chapter 5
Electronically Adjustable Current Mirrors

Abstract This chapter focuses on the design of an adjustable low-voltage current mirror. The presented topology provides continuous gain adjustment, while it simultaneously features the attractive characteristic of low-voltage operation. The behavior of the current mirror has been experimentally verified through a first-order lowpass filter fabricated in AMS 0.35 μm CMOS technology.

Keywords Analog integrated circuit design • Current mirrors • Programmable current mirror filters • Low-voltage circuits

5.1 Introduction

In CMOS analog circuit design, continuously programmable scaling of current signals is often required. For instance, it is used in auto-zeroed amplifiers, electronically programmable analog filters, circuits for biomedical applications as well as image processing tasks. Current mirrors with adjustable gain are widely used in these applications. Thus, this chapter focuses on the case of electronically programmable filters, where as it will be demonstrated in the following sections, the electronic adjustment of current mirrors' gain results in the programmability of the overall filter gain as well as its cutoff frequency. In that way, any effect from process, voltage and temperature (PVT) variations can be eliminated.

For this purpose, a number of adjustable current mirror topologies have been already introduced in the literature [1–6]. However, all these realizations have at least one of the following drawbacks:

- Absence of low-voltage operation capability [3]
- Limited range of gain adjustment [5]
- Employment of PMOS transistors for handling ac signals [[1, 3, 4, 6]
- Limited bandwidth and/or several restrictions imposed by the utilization of MOS transistors operating in weak inversion or triode region [4, 6]
- Requirement of a specific nonlinear form of input currents [2]

C. Laoudias and C. Psychalinos, *Integrated Filters for Short Range Wireless and Biomedical Applications*, SpringerBriefs in Electrical and Computer Engineering, DOI 10.1007/978-1-4614-0260-2_5, © Springer Science+Business Media, LLC 2012

In contrast with previous implementations, the proposed scheme of current mirror simultaneously offers low-voltage operation capability employment of only NMOS transistors biased in strong inversion, absence of any imposed restrictions concerning the input current and bidirectional adjustment of gain.

5.2 Current Mirror with Adjustable Gain

The scheme of the introduced programmable gain current mirror is shown in Fig. 5.1 [7]. This topology will be denoted in the next as Low-Voltage Adjustable Current Mirror (LVACM). The LVACM utilizes a conventional fixed gain low-voltage cascode current mirror, known as Flipped Voltage Follower Current Sensor (FVFCS) [8]. Note that, the current gain of FVFCS could not be electronically adjusted due to the fact that it is determined by the aspect ratios of transistors M_{n5}–M_{n6}. Performing a simple inspection of the LVACM, it is easily derived that the minimum supply voltage requirement is equal to $V_{th} + 2V_{DS,sat}$, where V_{th} and $V_{DS,sat}$ are the threshold voltage and drain-source saturation voltage of MOS transistor. Thus, the configuration in Fig. 5.1 is capable for operation in a low-voltage environment.

The circuit operates as a gain programmable current mirror with continuously adjustable current gain A_i. This is achieved as follows:

The input current i_{in} is transformed into a voltage $v_{in} = i_{in}/g_{m1}$ through the diode connected transistor M_{n1}, where g_{m1} is the transconductance parameter of transistor M_{n1}. Note also that the input resistance of the LVACM is equal to $1/g_{m1}$. This voltage is converted back to a current through the transconductance cell formed by transistor M_{n2}. Thus, the current that is fed into the FVFCS is equal to $(g_{m2}/g_{m1})i_{in}$,

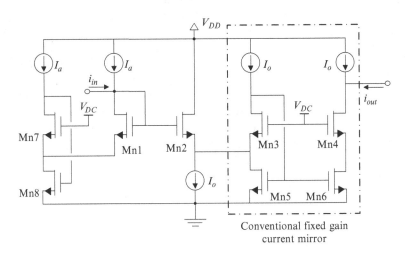

Fig. 5.1 Topology of low-voltage current mirror with programmable gain

where g_{m2} is the transconductance parameter of transistor M_{n2}. Neglecting the body effect and other second-order effects, the output current i_{out} can be expressed as

$$i_{out} = \frac{g_{m2}}{g_{m1}} i_{in} \tag{5.1}$$

Transistor M_{n1} is biased at a dc current I_a, while M_{n2} is biased at a dc current I_o. Taking into account that the small-signal transconductance parameter (g_m) is given by the formula $g_m = \sqrt{2K\frac{W}{L}I_0}$, where $K = \mu_n C_{ox}$ is the transconductance factor of MOS transistor, the current gain (A_i) of the LVACM is given by

$$A_i \equiv \frac{i_{out}}{i_{in}} = \sqrt{\frac{I_0}{I_a}} \tag{5.2}$$

From (5.2), it is revealed that the current gain of the current mirror can now be adjusted through the dc currents I_a and I_o. Moreover, the current gain of the FVFCS remains unaffected during the process of electronic adjustment. This is due to the fact that the gain of FVFCS could only be changed during the layout design by modifying the aspect ratios of transistors M_{n5}–M_{n6}.

To demonstrate the benefits offered by the LVACM, a simple CMOS current-mode lossy integrator has been designed. This has been achieved by connecting a capacitor between the common gate connection of transistors M_{n1}–M_{n2} and ground. The realized first-order filter function is expressed by

$$\frac{i_{out}}{i_{in}} = A_i \frac{1}{1 + \tau s} \tag{5.3}$$

In (5.3), A_i is the current gain factor given in (5.2), while the time constant is given by the formula $\tau = C/g_{m1}$. Thus, the low-frequency gain of the filter is determined by the current gain of the LVACM which is given by the formula $\sqrt{\frac{I_0}{I_a}}$, while the cutoff frequency is proportional to the factor $\sqrt{I_a}$. As a result, the low-frequency gain could be independently adjusted through the dc current I_o without disturbing the cutoff frequency which could be adjusted through the dc current. The overall topology of the first-order lowpass filter using the LVACM, where ideal current sources are replaced by transistors, and its symbol are shown in Fig. 5.2.

5.3 Experimental Results

The current-mode lossy integrator using the LVACM was fabricated through AMS 0.35 μm CMOS n-well process with NMOS and PMOS threshold voltages of about 0.5 and −0.7 V, respectively. Considering dc bias currents $I_o = I_a = 6$ μA, supply voltages $V_{DD} = 1.5$ V and $V_{DC} = 1$ V, the aspect ratios were 4/0.8 μm for

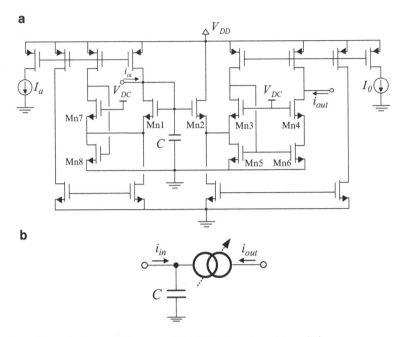

Fig. 5.2 First-order lowpass filter using LVACM (**a**) topology (**b**) symbol

transistors M_{n1-5} and 7/1 μm for transistors M_{n6-8}. The required dc current sources have been realized by PMOS transistors with aspect ratio 15/1 μm. Due to the fact that the input resistance of the LVACM was 8.75 kΩ, the value of the capacitor was chosen to be equal to $C_1 = 10.7$ pF, in order to achieve a cutoff frequency 1.7 MHz.

The input V/I and output I/V conversions play an important role in all current-mode signal processing systems, since the performance of the topology under test depends highly on the interfaces' performance. For this purpose, the feed of the current-mode integrator has been achieved by employing a typical V/I converter constructed from appropriately configured MOS current mirrors. The output I/V conversion is utilized by an appropriately configured on-chip transresistance amplifier. Finally, the reference currents I_0, I_a that are used to adjust the gain of LVACM are constructed by means of a low-voltage cascode current mirror and the external resistor R_{b1} and R_{b2}, respectively.

The employed general measurement setup including the off-chip components is demonstrated in Fig. 5.3. The die microphotograph of the fabricated chip is shown in Fig. 5.4, while the photograph of the PCB that was used for the experimental verification is depicted in Fig. 5.5. The time-domain behaviour of the first-order lowpass filter is demonstrated in Fig. 5.6, where the output currents are obtained by applying a 10 kHz sine-wave input signal with an amplitude 2 μA. The dc current I_o was swept from 4 to 8 μA with a step of 1 μA, while I_a was kept constant at 6 μA.

The capability of adjusting the low-frequency gain without disturbing the cut-off frequency is demonstrated in Fig. 5.7, where I_o was swept from 4 to 8 μA and I_a was

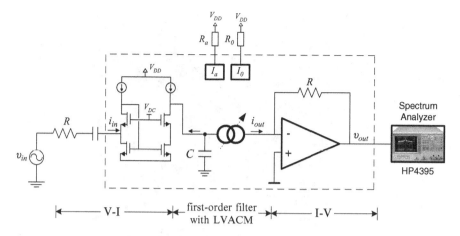

Fig. 5.3 Measurement setup for the first-order lowpass filter

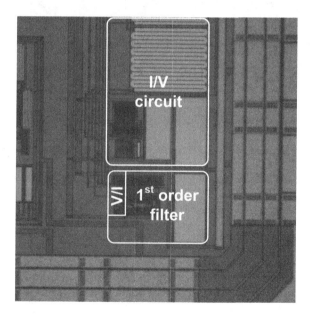

Fig. 5.4 Die microphotograph of the fabricated chip

held constant at 6 μA. The measured values of the gain were 0.7, 1, and 1.2, while the theoretically predicted values were 0.8, 1, and 1.15, respectively. The measured cutoff frequency in all the above cases was 1.708 MHz which is also close to that theoretically predicted. The capability for varying the cutoff frequency without disturbing the gain through a simultaneous adjustment of the dc currents I_o and I_a, has been experimentally verified as it is shown in Fig. 5.8.

Fig. 5.5 Photograph of the PCB used to test the fabricated chip

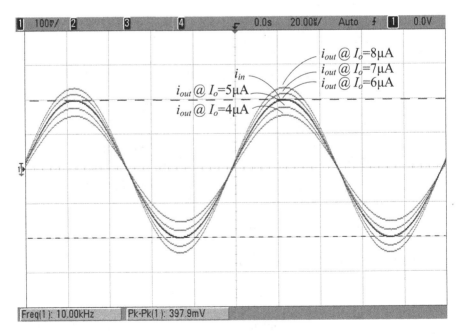

Fig. 5.6 Experimentally measured output current waveforms for $I_\alpha = 6\,\mu A$ and I_0 swept from 4 to 8 μA in 1 μA steps

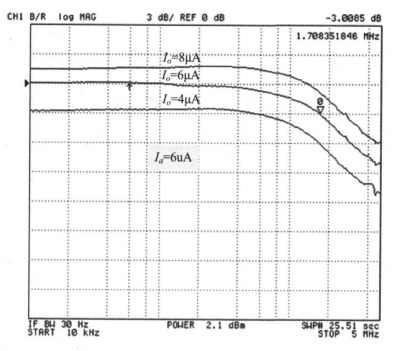

Fig. 5.7 Demonstration of electronic adjustment of low-frequency gain for $I_\alpha = 6\,\mu A$ and $I_0 = 4$, 6 and 8 μA

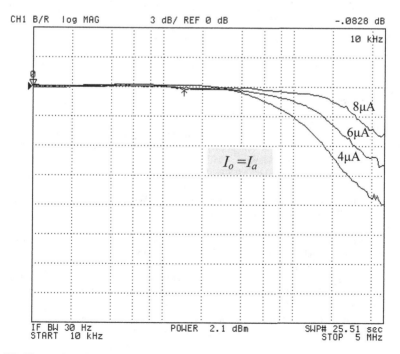

Fig. 5.8 Electronic adjustment of cutoff frequency without affecting low-frequency gain for $I_\alpha = I_0 = 4, 6$ and 8 μA

With regards to the non-linear distortion produced by LVACM, it depends on the values of the quiescent drain currents. This is originated from the fact that the gain depends on small-signal transconductances determined by the operation point of MOS transistors. This has been verified through experimental results, where the measured values of total harmonic distortion for a sine-wave input signal with 2 μA amplitude and 10 kHz frequency were 0.45%, 0.97%, and 1.4% at gain equal to 1, 0.7, and 1.2, respectively.

Finally, the silicon area of the current-mode lossy integrator based on the LVACM is approximately 86 μm $\times$ 74 μm, and the dc power consumption is 61 μW.

5.4   Summary

In this chapter, a novel topology of current mirror with adjustable gain and capability of low-voltage operation is presented. At first, a review on previous implementations that have been reported in the literature is performed, highlighting the drawbacks and restrictions of the corresponding circuits. Then, the operation of the proposed current mirror is studied, while it is used to design a first-order lowpass filter. Through this design example, the attractive characteristic of independent adjustment of the overall gain and the cutoff frequency is pointed out. Finally, the performance of the LVACM has been evaluated through experimental results, where both the time-domain and frequency response have been considered. Measurement results demonstrate the potential of LVACM to be a reliable building block for realizing modern high-performance analog processing systems.

References

1. Z. Wang, and W. Guggenbuhl, "Adjustable Bidirectional MOS Current Mirror/Amplifier", *Electronics Letters*, vol. 25, no. 10, pp. 673–675, May 1989.
2. E.A.M. Klumperink, and E. Seevinck, "MOS Current Gain Cells with Electronically Variable Gain and Constant Bandwidth", *IEEE Journal of Solid-State Circuits*, vol. 24, no. 5, pp. 1465–1468, Oct. 1989.
3. A.K. Gupta, J.W. Haslett, and F.N. Trofimenkoff, "A Wide Dynamic Range Continuously Adjustable CMOS Current Mirror", *IEEE Journal of Solid-State Circuits*, vol. 31, no. 8, pp. 1208–1213, Aug. 1996.
4. J. Ramirez-Angulo, C. Durbha, A.J. Lopez-Martin, and R.G. Carvajal, "Highly Linear Wide Tuning Range CMOS Transconductor Operating in Moderate Inversion", in *Proc. IEEE International Symposium on Circuits and Systems (ISCAS)*, pp. 805–808, May 2004.
5. J. Ramirez-Angulo, S.R.S. Garimella, A.J. Lopez-Martin, and R.G. Carvajal, "Gain Programmable Current Mirrors Based on Current Steering", *Electronics Letters*, vol. 42, no. 10, pp. 559–560, May 2006.
6. A. Zeki, and A. Toker, "Tunable Linear CMOS Current Mirror", *Analog Integrated Circuits and Signal Processing*, vol. 50, no. 3, pp. 261–269, Mar. 2007.

7. C. Laoudias, C. Psychalinos, "Low-voltage CMOS adjustable current mirror", *Electronics Letters*, vol. 46, no. 2, pp. 124–126, Jan. 2010.
8. R.G. Carvajal, J. Ramirez-Angulo, A. Torralba, J.A.G. Galan, A. Carlosena and F.M. Chavero, "The Flipped Voltage Follower: A useful Cell for Low-Voltage Low-Power Circuit Design", *IEEE Transactions on Circuits and Systems I*, vol. 52, no. 7, pp. 1276–1291, Jul. 2005.

Chapter 6
Conclusions

This book describes the design of low-voltage analog integrated filters using current mirrors, one of the most common building blocks in analog and mixed-signal VLSI circuits. Its purpose is to contribute in the existing research area of the design of analog filters using low-voltage current mirrors. In this direction, all the derived filters are constructed by the low-voltage high-swing current mirror, offering the advantages of low-voltage operation, absence of passive resistors and electronic adjustment capability of their frequency characteristics. Several design examples have been described in detail, namely universal biquad filter topologies, complex filters for Bluetooth/ZigBee low-IF receivers, wavelet filters for cardiac signal detection and a current mirror topology with electronically adjustable gain. Experimental results from the fabricated chips have been also presented, exhibiting their utility in modern low-voltage low-power portable devices.

In Chap. 2, novel SIMO and MISO universal biquad topologies have been presented. The experimental results from the fabricated chip verified the correct operation of both filters as well as the optimum layout design of current mirrors and capacitors. The chip prototype was fabricated through the AMS C35B4C3 0.35 μm n-well CMOS process. With regards to the dynamic range of the two filters, the measured SFDR was 42.7 dB for SIMO and 41.6 dB for MISO topology, being in good agreement to the simulated results of 44 and 42.6 dB, respectively.

In Chap. 3, the design of low-voltage complex filters using current mirrors for low-IF Bluetooth/ZigBee receivers has been studied. As a design example, two twelfth-order complex filters based on the leapfrog and topological emulation of LC ladders methods have been fabricated through the AMS 0.35 μm process. The experimental results confirmed their correct operation, since both filters meet the specifications of Bluetooth standard. More specifically, the leapfrog complex filter has SFDR = 40.4 dB, IRR = 28.2 dBc and in-band group delay variation 0.56 μs, while the corresponding values for the topological complex filter are 39.2 dB, 27.9 dBc and 0.41 μs.

In Chap. 4, the design of novel topologies suitable for realizing wavelet filter functions has been discussed. More specifically, the proposed wavelet filters are implementing a fifth-order rational function, which approximate the first derivative

C. Laoudias and C. Psychalinos, *Integrated Filters for Short Range Wireless and Biomedical Applications*, SpringerBriefs in Electrical and Computer Engineering, DOI 10.1007/978-1-4614-0260-2_6, © Springer Science+Business Media, LLC 2012

of Gaussian function. For this purpose, two different design approaches have been employed, namely the well known FLF method and the emulation of an orthonormal ladder filter. The integrators of both filters are constructed by low-voltage current mirrors operating in subthreshold region, offering thus the benefit of ultra low-voltage (0.5 V) and low-power operation. The efficiency of the proposed filters has been verified through simulation results by employing TSMC 130 nm CMOS process. Comparison results have shown that the topology derived from the orthonormal ladder filter has better performance in terms of power, capacitor area, linearity and dynamic range than that achieved by the FLF wavelet topology.

Finally in Chap. 5, a novel scheme of an adjustable low-voltage CMOS current mirror is introduced. The behavior of the derived topology has been experimentally verified through a first-order lowpass filter fabricated in AMS 0.35 μm CMOS process. The experimental results demonstrated the efficiency and usefulness of the current mirror in the synthesis of high-order programmable filters.

6.1 Motivation for Future Work

Considering the low-voltage current mirror with adjustable gain LVACM presented in Chap. 6, a very interesting prospect is the development of new circuits, like the auto-zeroed amplifiers for automatic compensation of offsets. Additionally, due to its attractive feature it could be used for the design of current-mode AGC circuits. Another important application of the current mirror is the electronically programmable filters. More specifically, having already studied the design of a first-order filter, the next step is the synthesis of higher order analog filters utilizing the LVACM. Therefore, the derived filters will operate under low supply voltage, while presenting the orthogonal adjustment of cutoff frequency and low-frequency gain, by modifying the bias currents of the current mirror. This feature is extremely important in integrated circuits, due to the fact that any effect from process, voltage, temperature (PVT) variations and transistor mismatching can be eliminated.

An interesting area of research with particular importance is the use of current mirrors in biomedical applications, like in Chap. 5 where novel structures of wavelet filters for the detection and analysis of cardiac signals have been presented. The results from these simulations pointed out the usefulness of these filters in implantable pacemakers, where high efficiency and low-voltage operation is required. Also, an important feature of the proposed structures is their low power consumption. Therefore, wavelet filters using current mirrors can be employed in a wide range of implantable devices, such as hearing aids, cochlear implants and neural stimulators/detectors for the detection of EEG (Electroencephalogram) signals. Finally, all the derived circuits can be combined and integrated into nowadays complex systems for low-power short-range wireless communications and biomedical applications.

About the Authors

Costas Laoudias was born in Sparti Lakonias, Greece in 1984. He received the B.Sc. degree in Physics, the M.Sc. degree in Electronics and Computers and the Ph.D. in Analog IC design, all from Physics Department, University of Patras, Greece in 2005, 2007 and 2011, respectively. In June 2011, he joined Analogies S.A. as an Analog/Mixed-signal IC Design Engineer working mainly in the area of high-speed I/O circuit design. His main research interests include design of low-voltage analog integrated circuits for signal processing, ultra low-voltage continuous-time filters for biomedical applications, IQ filters for wireless receivers.

Costas Psychalinos received the B.Sc. degree in Physics and the Ph.D. degree in Electronics from the University of Patras, Greece, in 1986 and 1991, respectively. From 1993 to 1995, he worked as Post-Doctoral Researcher with the VLSI Design Laboratory at the University of Patras. From 1996 to 2000, he was an Adjunct Lecturer with the Department of Computer Engineering and Informatics at the University of Patras. From 2000 to 2004 he was an Assistant Professor with the Electronics Laboratory, Department of Physics, Aristotle University of Thessaloniki, Greece. From 2004 to 2009 he was an Assistant Professor and currently he is an Associate Professor with the Electronics Laboratory, Department of Physics, University of Patras, Greece. His research area is in the continuous and discrete-time analog filtering, including companding filters, sampled-data filters, current amplifier filters, CCII and CFOA filters, and in the development of low-voltage active blocks for analog signal processing. He also serves as a member of the Editorial Board of the Analog Integrated Circuits and Signal Processing Journal.

C. Laoudias and C. Psychalinos, *Integrated Filters for Short Range Wireless and Biomedical Applications*, SpringerBriefs in Electrical and Computer Engineering, DOI 10.1007/978-1-4614-0260-2, © Springer Science+Business Media, LLC 2012